Automatic Control of
Industrial Drives

Automatic Control of Industrial Drives

C. H. Pike, C.Eng., M.I.E.E.

London

Newnes–Butterworths

THE BUTTERWORTH GROUP

ENGLAND
Butterworth & Co (Publishers) Ltd
London: 88 Kingsway, WC2B 6AB

AUSTRALIA
Butterworth & Co (Australia) Ltd
Sydney: 586 Pacific Highway Chatswood, NSW 2067
Melbourne: 343 Little Collins Street, 3000
Brisbane: 240 Queen Street, 4000

CANADA
Butterworth & Co (Canada) Ltd
Toronto: 14 Curity Avenue, 374

NEW ZEALAND
Butterworth & Co (New Zealand) Ltd
Wellington: 26–28 Waring Taylor Street, 1
Auckland: 35 High Street, 1

SOUTH AFRICA
Butterworth & Co (South Africa) (Pty) Ltd
Durban: 152–154 Gale Street

First published in 1971 by
Newnes-Butterworths an imprint
of the Butterworth Group

© Butterworth & Co. (Publishers) Ltd., 1971

ISBN 0 408 00062 7

Filmset by Photoprint Plates Ltd., Rayleigh, Essex

Printed in England by Fletcher & Son Ltd., Norwich

Foreword

By K. K. Schwarz, M.A., C.Eng., F.I.E.E., Technical Director of Laurence, Scott & Electromotors Ltd.

It is a real pleasure to write a foreword to this wide ranging survey of industrial drive equipments and installations, providing a comprehensive introduction to the main features of electrical drives and their control systems. It is too often forgotten that the objective of the engineer must be to produce the most efficacious overall solution to a drive problem. Moreover in considering the factors that matter, first cost and running cost, performance, reliability, servicing etc., there must be a critical appreciation of the requirements of the final product, be it the tension in a steel strip or in the dough of a swiss roll. Nor must the problems of the operator be forgotten. This aspect and the fact that there is no universal solution to all problems are very clearly illustrated by the text. The best and most economic choice for ensuring a correctly controlled water level in a reservoir must be investigated no less fully than a rolling mill drive, noting that the problem of water hammer in the former has its counterpart in the problem of impact speed drop in the latter.

This book, covering such a wide selection of different methods of converting electrical energy and applying controlled electric power to solve practical drive problems, will be a welcome addition to the literature.

Preface

Compiled specifically for technicians, works engineers, electricians, electrical fitters, apprentices and students interested in the automatic control of motor-driven industrial machines and processing plant, this book is intended to serve as an introduction to the engineering of drive schemes comprising integrated combinations of motors, motor power regulators, control apparatus and devices.

The discussion covers basic control principles; control requirements for various classes of drive; the types of motors and power regulators available for different applications; control circuits and devices; and typical examples of drives based on a.c. motors and on generator-fed and rectifier-fed d.c. motors.

The drives considered range from those for simple single-motor stop/start duties to multi-motor variable-speed sectional drives. The simpler drives are considered fairly briefly and without reference to the normal equipment used only for motor starting and not for automatic control of the drive. Main emphasis is on the more complex drives where the motors are regulated precisely throughout a duty cycle or continuously as required for purposes such as maintaining the level of a liquid in a tank, the pressure in a pipeline, or to drive a length of material through a series of processing units while maintaining required linear speeds and tensions.

Much of the discussion is concerned with drives for steel mills and other multi-motor strip material processing plants which have complex control requirements and are therefore particularly interesting examples of advanced types of drives involving highly sophisticated engineering techniques.

Although typical of drive engineering practice during the past decade, some of the earlier equipment and installations described have now been outdated by later developments, but being still in service they are of continuing interest as examples of the principles involved and which are still relevant to the more recent schemes, e.g. those based on thyristor converters.

In general, it is assumed that the reader has a good knowledge of conventional electrical technology and does not want to be told what he knows already—or can look up in standard text books. For instance, no attempt is made to discuss the basic principles of certain electronic devices, which in many cases may simply be regarded as 'black boxes' performing a particular function in a control scheme. There are already plenty of books dealing specifically with the contents of the black boxes.

As the intention of the book is to present authoritative, practical and contemporary information, full advantage has been taken of the extensive literature on the subject, and the main sources of reference are included in the Bibliography.

For the information abstracted from their publications, thanks are due to the companies named in the text as responsible for particular equipment and installations described. Thanks are also due to Belmos Peebles Ltd, for permission to use diagrams and text from the book *A.C. Motor Control and Distribution Gear*, by R. T. Lythall, and to the Institution of Electrical Engineers for matter abstracted from the paper 'Thyristor-supplied tandem cold mill' by D. W. Draper and R. T. Goodridge (see Bibliography).

It should be noted that certain of the companies responsible for publications issued some time ago do not now exist as such owing to amalgamations into the organisation now known as The General Electric Co, Ltd, and previously as The General Electric & English Electric Companies Ltd. It is also possible that some companies named may have changed their titles by the time the book is published.

Thanks are due to Mr. K. K. Schwarz for his interest and support in the preparation of the manuscript and for his foreword.   C.H.P.

Contents

1 Basic Control Techniques   1

2 Drive Control Requirements   6

3 A.C. Motors and Power Regulators   21

4 D.C. Motors and Power Regulators   47

5 Control Devices and Circuits   66

6 A.C. Motor Drives   86

7 Generator-Fed D.C. Motor Drives   102

8 Static-Converter-Fed D.C. Motor Drives   120

   Index   136

Chapter 1

Basic Control Techniques

An industrial electrical drive consists of one or more motors and the equipment for the manual or automatic regulation of the motor power supply for the purpose of meeting the control requirement for the driven unit. Adjustment of the motor power regulating equipment is initiated by manual or automatic control action exercised when it is necessary to regulate the operation of the driven unit.

Control Actions

The simplest control action is on/off switching to start and stop the driven unit. It can be effected manually or by an automatic controller receiving signals which cause it to initiate the appropriate switching operation.

With many drives, control action may be exercised continuously to achieve the required regulation of the driven unit. Consider, for example, a control requirement for the driven unit to operate at constant speed with varying load. The drive will consist of a variable-speed motor with switching and power regulating equipment which is used to vary the input to the motor and therefore its speed.

With manual control the operator regulates the speed according to the information he receives from a speed indicator, making an adjustment whenever the pointer moves perceptibly from the datum value he is trying to maintain. If the load alters gradually and the supply voltage is fairly steady, manual regulation can keep the speed constant within the limits set by the operator's ability to detect a deviation from the datum value. The amount of deviation that can be detected depends on the resolution of the speed indicator scale, the resolution being the smallest change which can be read.

Manual control may suffice for some drives but for more precise and continuous speed regulation the operator must be replaced by an automatic control system. This will include devices that enable the desired and actual speeds to be compared quantitatively to obtain an error signal that is used to initiate operation of the power regulator so as to minimise continuously the difference between the desired and actual speeds. To enable the speeds to be compared quantitatively, in an electrical control system they are represented by voltages obtained from the devices included in the basic d.c.-motor speed-control scheme shown in *Figure* 1.1.

Basic Speed-control Scheme

In this scheme the desired speed reference voltage (V_i) is set on a linear potentiometer connected to a constant-voltage source—represented by a battery. The voltage (V_o) representing the actual speed is obtained from a tachogenerator, i.e. a small generator with a permanent-magnet excitation, coupled mechanically to the motor so that its voltage output is proportional to the motor speed. As V_i and V_o are acting in opposition, any difference between the desired and actual speed values results in an error voltage signal (θ) into the amplifier, which gives an output for operating the power regulator. In this basic scheme speed regulation is effected by adjusting the power regulator to vary the voltage (V_a) applied to the motor armature.

It should be noted that a control system designed to respond rapidly to speed changes is energised continuously, and consequently there is always

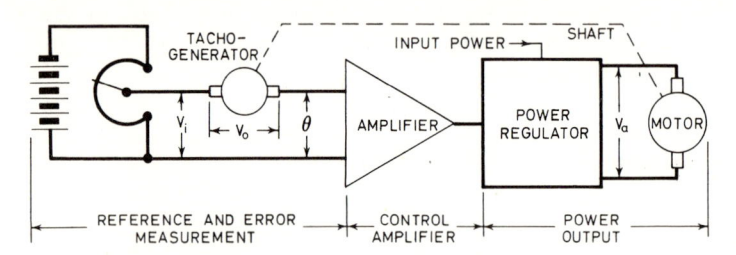

Figure 1.1. Basic d.c. shunt-motor speed-control scheme. The shunt field winding (not shown) is separately excited at constant voltage

some error voltage input to the amplifier. In practice, the desired and actual speeds always differ to some extent, owing to changing supply and load conditions, inaccuracies and losses in the system components and circuits.

A steady-state accuracy of $\pm 1\%$ of top speed is adequate for many drives but this can be improved to about $0\cdot2\%$ with advanced electronic control systems or, using a digital speed measuring scheme, to something within the order of $0\cdot02\%$ of top speed. These values apply only when the drive is running under steady conditions and the control system has only to correct relatively gradual changes in speed. If the drive is subjected to a severe disturbance due to excessive fluctuation of supply or load conditions, then the transient error may amount to 20% and the transient response time 1–2 s.

Function of Control Amplifier

An amplifier is used in a control system to increase its accuracy and also to provide for mixing signals from devices performing control functions in addition to speed regulation.

Although an amplifier gives an output for controlling a power regulator, this output is derived from the electricity supply energising the amplifier—it is not the amplified input signal. The function of the low-level input signal is to control the amplifying devices so that they regulate the electrical input to the circuits such that the output follows variations of the input signal. (This point is made because control-scheme diagrams do not usually show supply circuits to control units, unless they are relevant.)

An amplifier is not always necessary in a control scheme. For instance, an on/off scheme may only require switches actuated by a change in the controlled quantity, e.g. pressure, temperature, liquid level, and directly controlling the operation of the motor switching equipment.

With an amplifier a very small error voltage input gives an output at the level required for the control of a power regulator, consequently control action can be initiated by a relatively minute deviation from the steady-state error voltage. If the error voltage had to be large enough to provide a control input directly into the power regulator then the error would be of the order of volts instead of a fraction of a volt. In practice, a typical steady-state error voltage input to the amplifier would increase with rising speed from almost zero to about 50 mV at top speed.

Speed Regulating Control Action

It should be noted that when a scheme such as that shown in *Figure* 1.1 is used to vary the speed of the motor, V_i is altered by adjusting the potentiometer to change the input to the amplifier, and therefore the input to the power regulator and its output voltage V_a to the armature. To raise the speed, for example, V_a is increased to provide the current for accelerating the armature to the higher speed.

Adjusting the potentiometer to raise the speed initially increases the error voltage to a value that gives a power regulator output providing for the additional torque required to accelerate the armature to the higher speed. As this speed is approached, the increasing voltage V_o reduces the output of the power regulator until it reaches the steady-state value for the higher speed.

Progressive reduction of the error voltage with rising speed also reduces the acceleration so that when it is necessary to minimise the time to reach a given speed, the control scheme is designed to maintain a maximum permissible acceleration until the speed is nearly at its final value. The error voltage is then reduced to the appropriate steady-state value.

Current and Acceleration Limitation

If at the same time the drive has to be protected against the adverse effects of excessive current and rate of acceleration, the scheme is designed to limit these to predetermined values. Such safeguards are essential when the control potentiometer can be set to obtain a substantial rise in speed with a consequent initial error voltage which would cause the power regulator to supply excessive current to the armature.

To restrict the current to an acceptable value the error voltage can be reduced by a current-limit signal voltage into the amplifier. Alternatively, the control scheme can be designed for the automatic adjustment of the speed reference voltage to keep the error voltage at an acceptable value.

Closed-loop Systems

The scheme shown in *Figure* 1.1 forms a closed-loop system which can be defined generally as one that automatically corrects deviations from a desired value of a particular controlled quantity e.g. speed, torque, pressure or temperature. In *Figure* 1.1 the closed loop is the circuit used both for the control signal initiating operation of the power regulator to meet a speed requirement, and also for a feedback signal which resets the error voltage to the value appropriate to the speed requirement. The tachogenerator provides the feedback signal—also known as the reset signal.

An open-loop system is one in which there is no continuous feedback signal, representing the actual value of the controlled quantity, for comparison with a desired value signal to initiate automatic control action. However, if an open-loop system is controlled intermittently by an operator, in effect it becomes a closed-loop system whenever the operator reads the immediate value of the controlled quantity from an instrument to decide whether control action is necessary and acts accordingly. The intervention of the operator temporarily closes the control loop.

Control-signal Utilisation

Amplifiers are essential components of complex control schemes with various low-level signal inputs which have to be mixed. In effect, the amplifier computes from a mixture of signals the resultant output signal which, by controlling the power regulator, will at any instant meet the control requirements for the operation of the drive. These requirements are determined primarily by the need to ensure the proper functioning of the driven unit having regard to the purpose for which it is being used. For example, in machines processing strip materials it is essential to maintain a given tension in the strip as it is moved by some arrangement of two individual drives running at the predetermined relative speeds necessary to apply the tension. In this case, the control scheme must maintain the correct speed differential and at the same time the error voltage determining the speed setting is subject to modification by a signal voltage representing the strip tension. With any variation in the speeds of the drive, and during acceleration and retardation, the tension signal exercises overriding control in order to prevent any alteration in the tension, which would spoil a considerable length of the material being processed.

Although the drive is controlled to maintain the value of the controlled quantity within acceptable limits, the scheme also has to provide signals controlling the operation of the motor to enable its design performance to to be fully realised but not exceeded to an extent likely to cause a failure.

The motor control signals are mixed with the other inputs to the amplifier and if necessary act to modify the controlled quantity signal when it is calling for an operating condition detrimental to the motor. For instance, when strip tension is the controlled quantity, the required tension reference voltage is set on a potentiometer and compared with a feedback signal voltage representing the actual measured tension value. This comparison gives a steady-state error voltage used to establish the slight difference in speed between two drive motors necessary to apply the tension. Any deviation from the steady-state error voltage initiates adjustment of the speed of the motor which applies the tension so as to maintain the correct speed differential, and therefore the tension required.

If the strip breaks, the measured tension is zero, and consequently the feedback signal calls for an increase in tension which would mean that the motor applying the tension would begin to accelerate up to a speed determined by the error voltage, which is initially the tension reference voltage set on the potentiometer as at this instant there is no opposing voltage signal. To prevent the motor over-speeding when the strip breaks a signal is fed into the amplifier to reduce the error voltage to a value limiting the rise in speed and therefore protecting the motor.

Control-signal Sources

Various devices are used to sense changes in the value of the controlled quantity and give an output in a form that can be measured to provide a control signal. Some devices produce a voltage directly proportional to the controlled value, which therefore serves as the control signal to the amplifier.

One type of tachogenerator, for example, senses changes in speed in terms of volts because, being a generator with permanent-magnet excitation, its e.m.f. is proportional to its rotational speed. It can be used to energise a speed indicator, which is simply a voltmeter calibrated in revolutions per minute. Used for temperature control, thermocouples and resistance thermometers also produce a low-level voltage signal input to the amplifier.

Other sensors function by producing a mechanical motion when there is a change in the controlled quantity so that a transducer is also required to convert the motion into an electrical signal. A typical example is a pressure sensor with which a mechanical motion due to a change in pressure can be made to adjust a potentiometer giving a voltage output proportional to the value of the pressure being controlled.

Apart from the devices already considered, many others are involved in the practical control schemes used in conjunction with different types of motors and power regulators for automatic industrial drives. However, although some schemes are complicated by the inclusion of a multitude of control devices, in general the control techniques are based on the principles discussed briefly in this introductory chapter and applied in the specific schemes described in other chapters.

Chapter 2

Drive Control Requirements

Automatic controls are applied to industrial drives in order to achieve efficient operation of the motor-powered units with the minimum of human supervision. In effect, operation of the plant units is supervised by the control system which initiates the regulation of the drive motors necessary to meet the specified control requirements.

Start/stop Duties

For the on/off switching of a motor to start and stop the driven unit, the control scheme includes devices that sense when the motor must be started or stopped. For example, a float switch sensing when the level of a liquid has fallen to a predetermined datum will initiate the starting of a pump motor to raise the level to another datum at which the float switch initiates the stopping of the pump motor.

On/off switching for simple starting and stopping is effected by the usual motor control gear which may be required to restrict the starting current to protect the motor and/or control the slow acceleration of the driven machine. When it is not necessary to stop the machine quickly, the control gear is used to switch off the motor and the drive decelerates slowly.

Duty Cycle Requirements

With certain drives the starting and stopping periods form part of a recurring duty cycle, and since they are non-productive periods, the control requirement is that they should be minimised by accelerating and decelerating the drive motor at the highest rates practicable. In practice, although the drive may be designed for high accelerations, the rate will be limited to avoid excessive motor current and mechanical stresses on the driven unit. The control scheme will therefore include a current-limit feature which ensures that a predetermined maximum rate of acceleration cannot be exceeded. If the operating conditions are such that the motor could accelerate at an excessive rate when the load is not enough to make the current-limit control effective, then the control scheme will include a feature establishing a maximum permissible speed for every instant of the run-up period. It follows that by a predetermined regulation of the increase in speed with respect to time, the rate of acceleration is also regulated.

Control of the rate of acceleration also controls the rate of change of current which with d.c. motors must be limited in order to avoid commutation effects that would cause severe sparking. It should be noted that d.c. motors are used for industrial drives requiring the highest practicable rate of acceleration during run-up under load to the operating speed. In such cases, the rate of change of current can be the most onerous commutating condition.

A requirement for rapid acceleration to operating speed is usually accompanied by one for equally rapid deceleration to standstill. If this is effected by dynamic or regenerative braking of the motor, the rate of deceleration may have to be restricted by current-limit and speed/time control as applied during the run-up period.

Since the high rates of acceleration and deceleration are required to maximise the work done during a duty cycle, for the same reason the automatic control scheme must maintain the operating speed of the driven machine at the desired value.

If the duty cycle involves reversal of the drive, this must also be accomplished in the minimum practicable time.

It should be noted that the control scheme is required to achieve optimum values of rates of acceleration and deceleration; operating speeds; and reversal times. These optimum values are those which will ensure the most favourable operating conditions for the machine, and therefore its most economically efficient utilisation.

The ultimate criterion of efficiency of a machine is the cost per unit produced, and as this is influenced by capital and operating costs it follows that there is no point in attempting to increase productivity per duty cycle by pushing speeds up to values that give rise to excessive maintenance costs.

When the run-up and run-down times form a relatively high proportion of the duty cycle, designing a drive for the optimum rates of acceleration and deceleration contributes towards maximising the output of the driven machine. If several drive schemes are available for a particular machine, the one that shows a saving of only a few seconds in the run-up and run-down times may prove to be the most economical in the long term even if the initial cost is the highest.

Apart from ensuring optimum speed at every instant of a duty cycle, to minimise duty cycle time an automatic control scheme and the drive must respond with the minimum practicable time delay to signals calling for a change in speed.

Mine Winder Duty Cycle

The results obtained by the application of an automatic control scheme engineered to minimise duty cycle time are shown in *Figure* 2.1. The duty cycle illustrated is that of a mine-cage winding engine driven by the two

1070 kW (1440 hp) 600 V d.c. motors of a Ward-Leonard system. Regulated by a control scheme designed to minimise the time per wind, the winder raises a 12 t payload from a depth of 925 m (3025 ft) and has an output of 400 t/h. To achieve the optimum rope speed at every instant in the winding cycle, the control scheme ensures that the d.c. motor speed over the range zero to maximum is largely independent of variations in load, temperature and supply voltage and frequency; that acceleration and retardation are maintained at optimum rates independently of the load; and that the response to speed signal changes is rapid and accurate.

It will be noted from *Figure* 2.1 that after a 15 s acceleration period full speed is maintained for 60 s, then the 15 s retardation period is followed by a 3 s period during which the control scheme reduces the motor speed to

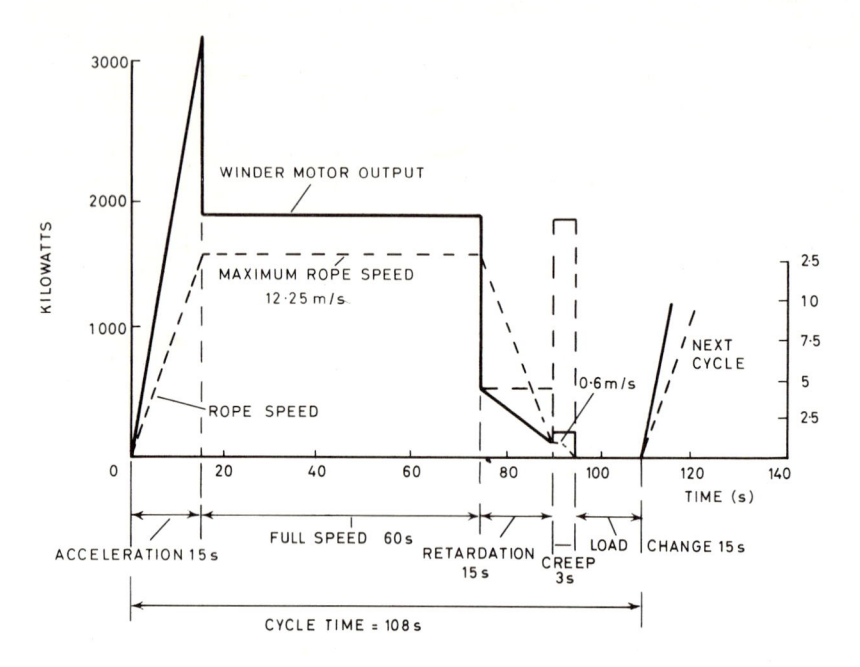

Figure 2.1. Power/time and velocity/time diagrams for the main coal-raising duty of a mine winder

a creep value that minimises the settling time before the mechanical brakes are applied. The economic advantage of having a control scheme to optimise speed values is emphasised by the fact that an increase in the winding time of 4·5 s would reduce the winds per hour from 33·3 to 32, and reduce the hourly output by 12 t.

Reversing Hot Mill Drives

A reversing hot mill is another example of a processing machine with a drive which is controlled throughout the duty cycle to optimise speeds. This type of mill is used for reducing steel ingot, slab, bloom, bar or strip by a series of successive passes through a pair of rolls that are screwed-down progressively, approaching one another as the gap between them is adjusted in the intervals when the metal has passed through the rolls and the work table is about to be reversed for the next pass in the opposite direction.

Operating conditions are severe as the roll gap is set to obtain the maximum reduction per pass which is permitted having regard to the mechanical load limit of the mill, the metallurgical limit of the material and the required cross-sectional profile of the workpiece. The speed of the mill must be controlled to achieve the maximum rolling rate permitted by the torque capability and rating of the drive, and also to minimise inter-pass time.

As the metal can be worked only while its temperature remains high enough, rapid acceleration to rolling speed, and reversal, are essential to minimise the temperature loss. There are, however, limitations to the maximum pass speed of reversing hot mills. The entry conditions impose a limit on the speed of entry, and due to load impact as the metal enters the rolls, the speed falls. The exit speed is the maximum at which the metal can be reversed and re-entered at the appropriate speed in about the time required to screw-down the rolls to the next setting. These conditions, together with the length of the metal and the acceleration and deceleration of the drive, fix a maximum speed.

With the d.c. motor drives used for heavy-duty reversing hot mills, reversal from base speed in one direction to base speed in the opposite direction is achieved in about 1·5 s and, for a 2 : 1 speed range, a top-to-top speed reversal in 3 s. Base-to-base speed reversals of less than one second can be obtained but are of little value unless the screw-down, roller table and other auxiliary drives have a correspondingly high performance. To prevent excessive loading of the motor, and excessive rate of change of current, the motor accelerates and decelerates under current-limit control.

Process-line Reeler Drives

Automatic schemes are essential for the efficient regulation of the drives of machines for the continuous processing of strip materials such as those produced by the metal, paper, textile and plastics industries. One control requirement is that the speeds of the drive motors must be regulated to maintain the critical values of the tension in the material and its linear speed within close limits to ensure the quality of the product and avoid breakages.

A typical production line comprises one or more individually driven processing machines, a motor-driven reeler (or coiler) for winding the strip into a coil, and a pay-off reeler (or decoiler) from which the strip is unwound and fed into the processing line.

The principal requirement of a reeler is that while winding up the material leaving the production line, it must also keep the coil suitably tight by maintaining a predetermined tension in the strip.

There are two basic types of reeler. One is the surface-driven type, consisting of one or two driving drums, against the surface of which the coil of material is held and caused to rotate. The single- and two-drum units are usually driven by a single motor, but in some cases the latter has a motor for each drum as this permits adjustment of the tension in the material between the two drums, and enables the optimum formation of the coil to be obtained.

With the centre-spindle reeler, the second type, the material is wound on to a spindle which is coupled directly to the motor. It has the advantage that the driving power is not applied to the surface of the material so the risk of damage is avoided. Equipped with two spindles (*Figure* 2.2), the reeler is used on processing lines where the material must be kept moving. With the two spindles, when one reel is full, the material coming from the line is wrapped around the empty spindle and the end of the full coil is cut through so that it can be removed from the reeler. This type of reeler is often used in the paper and plastics industries.

Each spindle of the reeler shown in *Figure* 2.2 has its own driving motor but only one is running until when the upper spindle is nearly full, the other motor is started and run up to such a speed that the peripheral speed of the empty spindle is equal to the linear speed of the material. At the same time the turret is rotated in an anti-clockwise direction until the empty spindle is in contact with the material (*Figure* 2.2). The turret arm then stops in this position and a serrated knife operates to sever the material from

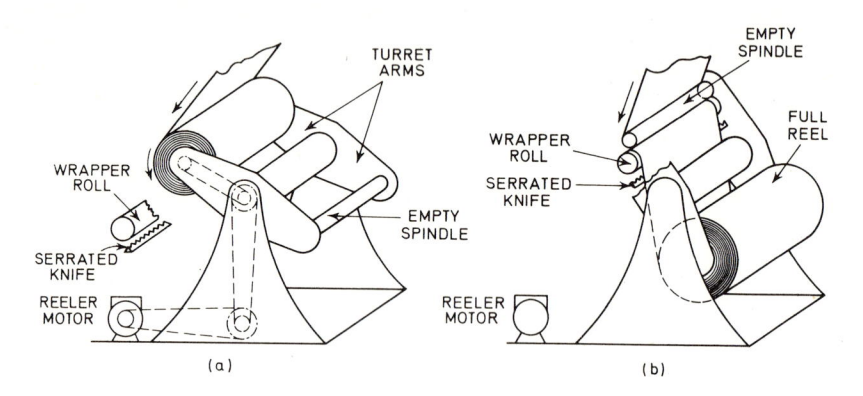

Figure 2.2(a). Typical centre-spindle reeler stand in normal reeling position (b) Position of reeler turret arms immediately before changeover from one spindle drive to the other

the full reel while the free end of the material is being secured to the empty core by the action of a wrapping roll. The full reel is removed while reeling continues on the second spindle.

The two-spindle reeler is also used for unwinding material being fed into a processing line which continues to operate while the edge of a full reel of material is being attached to the end of the length of material being processed. To make this 'flying splice' with paper and plastics materials, adhesive is applied to the leading edge of the material on the full reel which is then run up to speed. At the same time the turret is rotated to bring the leading edge into contact with the end of the material being processed so that continuity of the strip is maintained. The tail end of the material on the empty reel is severed.

The splicing of metal strip being processed continuously is effected by welding, and to avoid stopping the line while the joint is being made a reserve length of strip is held as a series of loops at the entry to the processing sections of the line (see page 128).

Reeler Control Systems

The control of both types of reeler can be effected by either a free reeler or a tension reeler system.

With the first, the reeler motor speed is controlled to obtain the required linear speed of the material, but the tension applied to the material results from it being wound up against a retarding effort imposed by some device (e.g. a mechanical brake) at the end of the processing line. The retarding effort may also be imposed by a drag generator (see page 13) being rotated as the material is wound up. It will be noted that the tension depends only on the retarding effort and is not affected by variations in the speed of the reel.

With the tension reeler system, the reeler may have to maintain the tension applied during processing or may be required to wind at a lower or higher tension than that applied in the preceding processing section. When the reeler has to maintain the correct processing tension in conjunction with a preceding drive motor, this is effected by regulating the torque of the reeler motor and the motor speed is reduced, as the coil builds up, to keep the peripheral speed and the linear speed of the material constant through the processing section.

To increase or decrease the tension applied in the processing section, the reeler motor winds the material at a linear speed higher or lower than that at which it is leaving the section. This exit speed is set by a drive motor preceding the reeler.

Assuming that the drive motor and the reeler motor are running initially at speeds that maintain the same linear speed from the processing section exit to the reeler, tension is increased by accelerating the reeler to a higher speed thereby stretching the material. When the required tension is achieved the reeler motor speed is held at the value that takes into account the increase in the linear speed of the material due to its elongation.

To reduce the tension, the reeler motor is decelerated to allow the material to relax and the speed is held at the appropriate lower value.

The free reeler motor also serves to pull the material through the processing sections when the operations involved do not require individual drives for each section. With the tension reeler the motor has to wind the material at the correct tension and does not control the line speed for this is effected by the drive (or drives) preceding the reeler. In general, the tension reeler is used in conjunction with one or more section drives and the speeds of all the motors have to be established and maintained at fairly precise values to ensure correct tensions and linear speeds throughout the line.

With the surface-driven reeler the motor speed must be maintained at a constant value to ensure a constant peripheral speed of the driving drum (or drums) and therefore constant linear speed and tension of the material. Being in contact with the drum as the reel winds up the material, the coil diameter increases and its rotational speed decreases but the linear speed of the material remains constant. To achieve constant tension therefore it is only necessary to keep the motor torque constant.

To maintain constant tension with centre-spindle reelers the motor speed must decrease progressively in direct proportion to the increase in coil diameter. If the motor speed remained constant with increasing diameter, the peripheral speed of the coil would increase and the tension would soon be excessive due to the material being stretched against the resistance exerted by the preceding drive running at constant speed. In some systems the line speed would also tend to increase. The control requirement is therefore for a scheme which responds continuously to the increasing coil diameter to give immediate and infinitely variable speed control of the reeler motor.

As an example of the speed-control requirement consider that in one revolution of the reeler spindle the strip material travels a distance equal to the coil diameter × 3·14. If the diameter is 1 m, the strip travels 3·14 m in one revolution so that with a spindle speed of 100 rev/min the strip speed is 314 m/min but when the diameter has built up to 2 m the speed must have fallen to 50 rev/min to maintain the same strip speed and tension.

As the motor has to slow down in direct proportion to the increase in coil diameter it has to provide constant power over a speed range corresponding to the minimum and maximum diameters of the coil. The ratio of these diameters is termed the build-up range.

The constant-power requirement results from the reeler motor having to develop increasing torque at decreasing rotational speed as the coil diameter builds up. Power is proportional to torque × speed (rev/min), and the torque developed by the motor is proportional to turning effort force × radius at which the force is effective. As the motor has to develop the torque to maintain rotation against the opposing torque exerted by the tensioned strip material at the radius of the coil, the motor torque has to increase to balance the increasing opposing torque as the coil diameter builds up.

The torque required from the motor at any instant is, therefore, proportional to strip tension × coil radius so that with tension reelers the control problem is to relate the torque of the motor to the coil radius. The problem is complicated by having to compensate for variations in mechanical losses and the effect of inertia during acceleration and deceleration of the reeler (see page 10).

The speed range required is a factor determining the rating of a centre-spindle reeler motor. For example, if the diameter of the empty spindle is 30 mm and the diameter of the full reel 90 mm, then the motor must be rated for constant power over a 3 : 1 speed range and the rating will be much greater than that of a motor for the corresponding constant speed surface-driven reeler.

With some continuous processing lines a taper tension control facility is used to obtain a progressive reduction in tension with increasing coil diameter for the purpose of preventing the centre of the coil becoming distorted by the pressure of the outer layers. It also prevents telescoping of the coil, i.e. a progressive movement of the outer layers to one side of the reel giving a dome-shaped appearance to the ends.

When a reeler forms part of a group drive, it has to be controlled independently to maintain the correct tension but to do so it must also follow closely any variations in the speed of the main drive. During acceleration and deceleration the tension in the material will alter unless the torque of the reeler motor is adjusted to compensate for the inertia of the reel, which is effective only during speed changes.

Although not always necessary, inertia compensation is required if there is a risk of variations in tension breaking the material or allowing it to become excessively slack. The inertia of large coils of low-tension material may present special problems when used on high-speed reelers, especially if frequent starting and stopping are required.

The amount of compensation required to maintain correct tension during speed changes is very critical, so that an essential feature of a reeler automatic control scheme is the circuitry which detects any tendency to a change in speed and initiates immediate regulation of the motor torque to minimise the variation in tension.

To achieve precise tension control, inertia compensation must be supplemented by compensation for variations in the mechanical losses of the reeler drive with changes in speed.

When the material is fed into the processing plant from an unwind reeler (also termed a pay-off reeler or decoiler), this must be controlled to allow the material to be pulled off at a speed determined by the process requirements, and the necessary tension is obtained by restraining the reeler.

Mechanical brakes have been used to apply this restraint but with the high-speed and powerful drives employed for many processes these are not practicable, owing to the difficulty of dissipating the heat produced and the cost of the electrical energy consumed by the main drive working against the brake. It is also difficult to provide the automatic regulation required to maintain constant tension as the coil diameter decreases, or to meet the other control requirements.

Unwind reelers are usually of the centre-spindle type and are coupled to a machine that functions normally as a drag or brake generator and may also have to function as a motor in order to pay out material during threading up prior to the production run. The machine may also have to motor during acceleration of the main drive in order to maintain the correct tension by assisting the material to unwind the coil. During deceleration of the line the restraint applied by the unwind machine must be controlled precisely to ensure that the correct tension is maintained.

The problem of maintaining constant tension with a centre-spindle unwind reeler is basically the same as that for the rewind reeler except that the machine is running normally as a generator and its speed has to increase as the coil diameter diminishes.

Reeler motor speeds have to be co-ordinated with the speeds of the other drive motors in order to maintain the required tension of the strip material through the one or more sections of the line. With centre-spindle reelers this co-ordination is complicated during acceleration and deceleration of the line because at the same time the diameter of the unwind coil is decreasing and the diameter of the rewind coil is increasing, consequently the reeler speeds are changing at the instant when changes in line speeds are initiated. Also, with the continuous variation in coil diameters, the reeler inertia is changing continuously so that the control system is required to adjust the inertia compensation continuously so that it is correct at any instant when a speed change may be made.

For some processes, the linear speed and tension of the strip may be maintained at the same values right along the line but if the material is elongated, either by processing or by the application of tension at one or more drive points of the line, the material will have to be kept moving at different linear speeds.

Two-stand Rolling Mill Drives

Consider, for example, the basic arrangement of a two-stand rolling mill (*Figure* 2.3) for reducing the thickness of metal strip. When the screwdown rolls are not set for reducing, the strip passing through stand 1 enters stand 2 at the same linear speed, and with identical motors driving each stand their

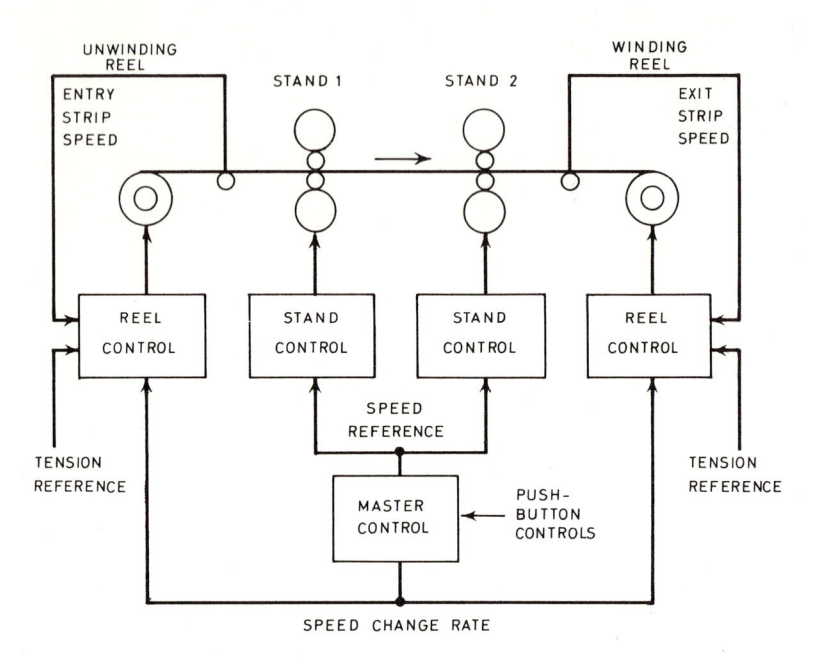

Figure 2.3. Basic arrangement of a two-stand cold mill and control system

speeds should be equal. If, however, the rolls of stand 1 are screwed-down, the strip is being reduced and elongated in the forward direction so that its linear speed is increasing.

With both stand motors running at the same speed, a loop would grow between the stands owing to elongation of the strip, and to prevent this stand 2 motor must run at a higher speed than stand 1 motor. In a cold rolling mill a required strip tension must be maintained so that any tendency to looping must be avoided, but with hot rolling a loop of material is often introduced between stands to ensure freedom from tension despite variations in temperature of the rolled stock.

The regulation of the drive speeds of the two-stand cold mill shown schematically in *Figure* 2.3 is effected by an overall control system consisting of master, stand and reel controls. The reel controls maintain the preset tension in the strip between the reels and the adjacent stands; the stand controls maintain the stand motors at preset speeds; and the master control provides overall simultaneous regulation of all the drive motors to ensure that their relative speeds are maintained during the complete rolling cycle of threading the strip, accelerating to and running at top speed, decelerating and ejecting the strip. The control system must also provide for any drive being jogged slowly in either the forward or reverse direction from standstill.

A typical cold mill can roll steel strip down to 0·1 mm (0·004 in) thick at speeds greater than 1524 m/min (5000 ft/min) using drives of 14 168 kW (19 000 hp) or more. The strip is threaded through the mill at a speed of about 61 m/min (200 ft/min) and when tensions have been established the mill is accelerated to top speed in about 5 s, the acceleration rate being about 305 m/min/s (1000 ft/min/s).

Reduction of the strip to the required gauge (thickness) depends on the predetermined rolling speed being maintained so that it is essential for the mill to be run up and down in the minimum time to ensure a minimum amount of off-gauge material. High rates of acceleration and retardation also help to maximise output since the total time taken to roll a complete coil is only in the region of five minutes. On high-speed tandem mills (i.e. multi-stand mills), acceleration and deceleration times may be about 5% of the total rolling time. Rapid retardation is also necessary for stopping the mill quickly in an emergency without breaking the strip.

Control of a cold mill is effected usually by push-button or switch selection of one of four operating conditions: 'stop', 'thread', 'hold' and 'run'. When 'thread' is selected the mill runs up automatically to a preset speed, which is usually in the range 15–150 m/min (50–500 ft/min), and when 'run' is selected, the mill accelerates to the preselected top speed. The selection of 'hold' at any instant during the run-up stops the acceleration and the speed is held indefinitely at the value it has reached. If then 'thread' is selected again, the mill speed falls at the normal retardation rate to the threading value. The selection of 'stop' above that speed causes the mill to decelerate rapidly under regenerative current conditions.

Where the stand power is above about 3700 kW ($\simeq$ 5000 hp), it becomes very difficult to transmit the total torque to the work rolls through one shaft and pinion box, so separate drives are used for each roll. This arrangement, known as a twin drive, involves co-ordinated control of the two motors with those of the other stands in a tandem cold mill.

With modern cold mills the large coil build-up ratios used necessitate a wide range of speed regulation of the reeler motors to obtain a required range of strip tension control. The range of strip tension obtainable electrically with one motor is approximately 10 : 1, but some mills require wider ranges, particularly single-stand mills where several passes are performed in the same stand. Tension ranges up to 50 : 1 can be obtained either by using two motors with a clutch between them, or by arranging for a change in gear ratio. Occasionally the two methods are combined. Overall tension ranges of more than 50 : 1 are difficult to achieve in practice, if accurate tension control is essential, because of the problem of compensating for the drive mechanical losses which fluctuate slightly in a random manner.

Reelers are usually fitted with holding brakes to stop the coil turning while it is being handled and, where the mill acceleration and stopping rates are limited by the reeler, disc brakes are used to minimise the drive inertia.

The control scheme for the screwdown motors of a cold mill provides for both screws being operated clutched together or, alternatively, independently in the same or opposite direction simultaneously, and the motors are required to respond rapidly to control signals. A position-control system may be used for the regulation of the operation of the screwdowns.

Automatic Gauge Control

Many users of metal strip, especially those pressing it to form products such as car bodies, require the strip thickness to be maintained to an accuracy of $\pm 1\%$ or less of the nominal value. To achieve this over the complete length of the coil, automatic gauge control is used since it can initiate corrective action much faster than an operator.

Variations in strip thickness can be attributed to two basic causes; first, irregular incoming gauge profile, second, changing mill behaviour due to speed and temperature changes or roll eccentricity. Even with incoming strip within the required $\pm 1\%$ tolerance, gauge control problems would still be experienced on the cold mill because variations in mill speed during acceleration from threading to top speed can reduce the outgoing strip speed by as much as 10%. However, there is often a large variation in the thickness or hardness of the incoming strip and, in addition, when large coils made up of two or more smaller coils welded together are used, there may be step changes in thickness of up to 8%

The two variables available for correcting gauge errors are rolling load and tension. An increase in rolling load by moving screws to bring the top roll down causes a reduction in strip thickness, and the same effect is achieved if either front or back tension is increased. Variation in tension of up to 20% of the set value is usually permitted except at the end of a mill where constant tension is necessary to form an even coil.

The simplest effective form of gauge control for a single-stand mill is a system using both tension and rolling load control. Small gauge variations are corrected by altering the back tension within permissible limits. Outside these limits the screws are adjusted and at the same time the tension is reset to its original value. For large gauge errors, such as occur at the beginning of a pass or at a weld, the screws are pulsed until the gauge error is within tolerance. The pulses are adjusted so that the screwdown 'on' time is a function of both the gauge error and of a preset hardness control, which takes account of the metallurgical hardness and cross-section of the material being rolled. The spacing between pulses is an inverse function of the mill speed, thereby allowing a change in strip gauge due to the movement of the screws to reach the measuring device in time to assess whether further action is necessary.

With a mill rolling foil, automatic gauge control (a.g.c.) differs from that used for thicker gauges in that control is effected principally by vernier control of the mill speed. Gauge is reduced by increasing speed and this method of correction is used instead of screwdown action.

Tandem mill a.g.c. is designed individually for each application so that it is most effective over the specified speed range for the metal rolled. There are several variations of combined screw and tension control which will provide a.g.c. *Figure* 2.4 shows a scheme for a three-stand tandem mill with which the stand roll forces are maintained constant and the a.g.c. adjusts the interstand tensions. These tensions are measured by interstand tensiometers and a.g.c. is initiated by three X-ray thickness gauges located at the delivery side of each stand.

The stand 1 gauge detects large rates of change of incoming thickness, such as would occur at welds, and thereby introduces an anticipatory control to enable the stand 2 system to make the correction before the strip reaches the stand 2 gauge. This gauge control operates on the speed of stand 1 to vary the tension between stands 1 and 2 within acceptable limits. The stand 3 gauge control provides a final vernier trim of thickness by varying the speed of stand 3 and therefore the tension between stands 2 and 3 within

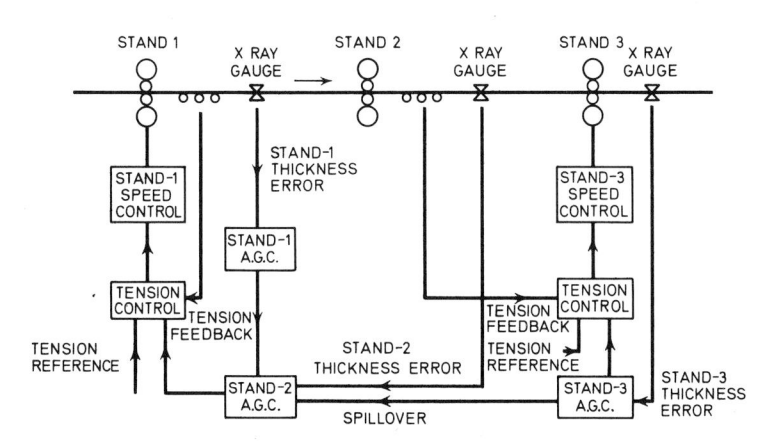

Figure 2.4. Automatic gauge control system for three-stand tandem mill

acceptable limits. If the gauge error requires a tension value outside these limits, the control signal passes through a spillover circuit to regulate the thickness out of stand 2 to a more suitable value.

Continuous Hot Mill Drives

The control requirements for a continuous hot mill differ from those of a cold mill owing to the special conditions complicating tension control. In recent years there has been a demand for higher throughput and for finer tolerances on such factors as rolled gauge, cut length and surface finish. Faster and more accurate control schemes have become necessary to achieve the required performance by minimising the effects of impact speed drop and temperature changes.

The drive motors and control scheme for a continuous hot mill must be designed to minimise impact speed drop. Using a d.c. motor, without automatic regulation the speed drop is in the region of 4–8% of the base speed and there is a recovery to a steady value of $1\frac{1}{2}$–2% as the armature current increases and the motor develops sufficient torque to meet the applied load.

The impact drop produces a loop of slack strip between the stand the material is entering and the preceding stand; this loop may be difficult to accommodate. Also, if the speed has not recovered to its final steady value by the time the strip enters the succeeding stand, excessive inter-stand tension may cause the end product to be off-gauge. To ensure freedom from tension, a loop may be used between stands and regulated by a control scheme to maintain the required loop depth.

The effect of impact speed drop is shown in *Figure* 2.5 which gives examples of curves obtained from analogue computer studies carried out by AEI

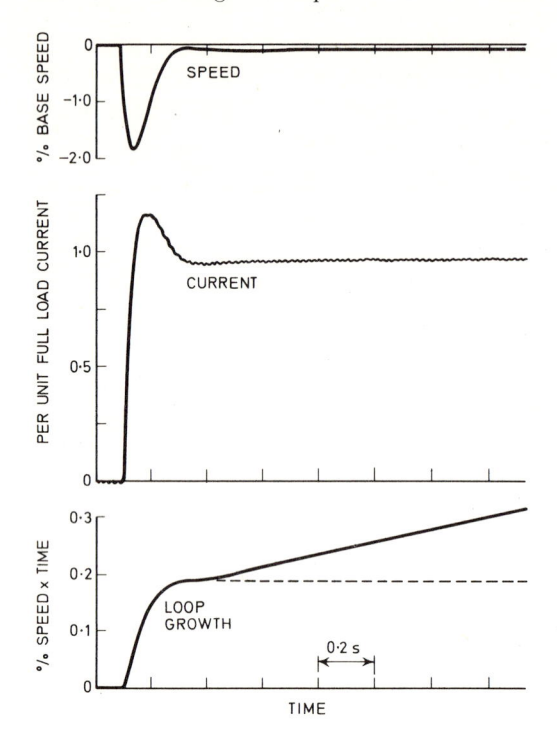

Figure 2.5. Curves showing the effect of impact speed drop on loop growth on a hot mill powered by a d.c. shunt motor with full-field excitation. The loop growth after impact is the integral of the steady-state error and is not relevant

during the design of a control system for a rod and bar mill equipped with individually driven speed regulated stands. With the system the steady-state speed regulation was 0·1%, and the time to return to speed coincidence after full-load impact at base speed was 0·2 s, and similarly at top speed.

The control scheme for a hot mill has to take into account the reduction in temperature of the material as it passes through the mill so that the back end becomes relatively cooler than the front as rolling proceeds. This effect is especially pronounced with mills rolling wide strip due to the large radiating surface area. In addition, the forward slip of the material increases as the rolling temperature decreases within the working temperature range. Owing to forward slip of the material between the rolls, the delivery speed from any given stand may be somewhat higher than the peripheral speed of the rolls. If the stand speeds are set up to take into account forward slip with given temperature conditions, any departure from these will affect loop growth. To minimise temperature effects, continuous hot mills are operated at the fastest practicable speed to maintain maximum temperatures throughout thereby achieving a high throughput of on-gauge material.

Process-line Sectional Drives

Metal rolling is essentially a batch process but subsequent treatments of strip are effected continuously by line set-ups with facilities for maintaining the supply of material, and removing coils of finished material without stopping the line.

The operation of continuous-strip processing lines involves the co-ordination of individual motor drives to maintain speeds and tensions within close limits. At the same time the output is maximised by operating the line at the highest speed practicable and this depends on how precisely the drives can be controlled to ensure the minimum deviation from the process parameters which must be maintained to obtain the required quality of the product.

The control requirements for continuous process lines are also determined by the nature of the product. With paper machines, for example, in the early stages of processing the paper web is stretched as the bulk of the moisture is removed but drying at a later stage causes shrinkage. Consequently, precise regulation of the relative speeds of the sectional drives is essential to maintain a constant weight of paper and to avoid breaking the weak web linking the drives. But although the control scheme must provide rapid response to minute deviations of speed signals, as the machine runs for a long time without stopping, acceleration and deceleration times are not important whereas they are a significant proportion of a batch process duty cycle which is accomplished in a few minutes.

With some production plants, units have to be started up and shut down automatically in a given sequence so that the control scheme involves interlocking of the motor control circuits to provide sequential switching with time delays between the starting of each drive. Sequential switching of conveyer system motors is used when the final conveyer delivering material to a process must be running before the conveyers which are feeding it with material.

Single-motor Drives

Automatic regulation of single-motor drives is used either to maintain a constant speed or to vary the speed as required for the operation of the driven unit. The control requirement for machine tools, for example, may be for the drive speed to alter slowly with the movement of the tool holder in order to maintain a constant cutting speed.

With increasing utilisation of automatic drives for a variety of industrial processes, there is a continuing development of schemes to meet specific requirements. Such schemes are based on the utilisation of the type of motor and power regulator which meet the control requirements determined by the function of the driven unit.

The required performance characteristics of a control scheme depend on the type of drive involved. In the interests of achieving the economical engineering of a drive the object is always to adopt the simplest form of scheme that will meet the control requirements. There is no purpose in using a scheme with performance characteristics superior to the inherent characteristics of the drive, e.g. a very rapid response scheme is not necessary if the variations in the controlled quantity occur slowly and/or it does not have to be maintained very close to a desired value.

Chapter 3

A.C. Motors and Power Regulators

Advanced forms of electrical drives have been developed for the latest high-performance industrial plants and machines but many drive requirements are still met by basically simple schemes. However, some such schemes have been improved by the incorporation of recently introduced electronic control components. For instance, the well known Ward-Leonard system invented 75 years ago is still used, although the performance of the basic scheme has been improved considerably by the inclusion of electronic units for certain control functions.

The combination of motor and power regulator selected for a particular industrial drive is that which, in conjunction with the control equipment, ensures the most effective utilisation of the driven unit.

Squirrel-cage Motor Schemes

The simplest drive for an automatic on/off control requirement is a squirrel-cage motor with a direct-on-line starter which stops and starts the driven unit when some switching device, e.g. a float switch, closes or opens the contactor control circuit. The starter serves as a power regulator by switching the supply on and off, which is all that is required for the operation of the driven unit.

For many start/stop duties the electro-mechanical contactor starter is entirely satisfactory but for applications involving high speed and frequent switching the thyristor contactor is available. Although the cost of the thyristor contactor is relatively high, it has the advantage of being a static device and therefore capable of a very long service life with negligible maintenance.

Thyristor Contactors

The thyristor contactor consists essentially of the thyristors and the pulse generators controlling them. In principle, the thyristor itself is a switched semiconductor diode. Referring to *Figure* 3.1, the arrowhead of the thyristor symbol is the anode (positive) and the line is the cathode (negative). By convention, the current flow is assumed to be from positive to negative (although electron movements are from cathode to anode) so that the arrowhead of the symbol indicates the direction in which a thyristor (and a diode) conducts easily.

The normal flow through a thyristor is termed forward current and it is maintained by the forward voltage. The voltage acting in the opposite

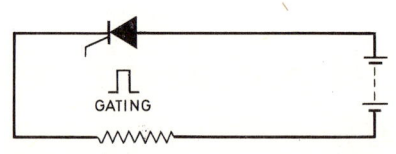

Figure 3.1. Operation of thyristor on d.c. supply

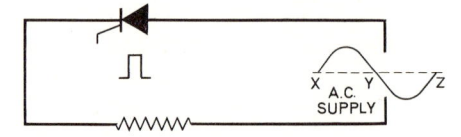

Figure 3.2. Operation of thyristor on a.c. supply

direction, i.e. that in which there is normally no appreciable current flow, is termed the reverse voltage. If this voltage exceeds a particular value it will cause a reverse current to leak through from the cathode to the anode.

In addition to its anode and cathode, a thyristor has a third electrode which is termed a gate. Like a diode a thyristor blocks reverse current but, unlike a diode, it also blocks forward current until it is turned on (i.e. switched on) by a gating pulse—a short low-voltage pulse applied between the gate and the cathode.

If a d.c. forward voltage is applied to a thyristor (*Figure* 3.1) which is then turned on, it continues to pass forward current indefinitely either until the forward voltage is removed or until a reverse voltage, greater than the forward voltage, is superimposed on the forward voltage; the gate cannot be used to turn the thyristor off. However, it is only necessary to remove the forward voltage, or apply a reverse voltage for a few microseconds to turn the device off.

As shown in *Figure* 3.2, when an alternating voltage is applied and the thyristor is turned on at instant X (when it is forward-biased), the thyristor conducts until the forward voltage falls to zero at instant Y. It cannot conduct again unless a further gating pulse is applied at instant Z. Between instants Y and Z the thyristor is reverse-biased and therefore blocks the circuit against reverse current.

If the thyristor is turned on at the instant when the supply voltage begins to apply a forward bias, it conducts only while the forward voltage persists; i.e. until the end of a single half-cycle. When the applied voltage reverses, the thyristor is reversed-biased, but when the applied voltage reverses again so that the thyristor is forward-biased for the second time, it will not conduct unless a gating pulse is applied.

A.C. Motors and Power Regulators

A thyristor can be made to conduct throughout every alternate half-cycle by applying a timed series of gating pulses. Since, however, a thyristor does not conduct until it is turned on, by delaying the application of the gating pulse it can be turned on at any desired instant during each forward-biased half-cycle. With this form of control, the thyristor can be made to conduct for only parts of half-cycles. The shorter the time during which the thyristor conducts in each half-cycle, the less is the average voltage during the half-cycle. Thus, by varying the instant at which the thyristor is turned on, the output voltage can be varied.

As a thyristor conducts only when a forward voltage is applied, a single-phase circuit with a thyristor in one line (*Figure* 3.2) gives a pulsating unidirectional output. To obtain an a.c. output a second thyristor is included—as shown in *Figure* 3.3. With this arrangement each thyristor conducts in turn when the forward voltage is in one direction or the other.

For 3-phase switching various arrangements of thyristors are used. *Figure* 3.4 shows a scheme applicable to motors with star- or delta-connected

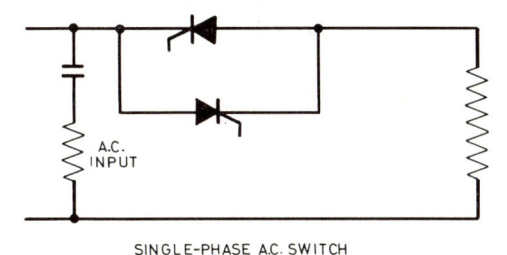

Figure 3.3. Single-phase a.c. thyristor switch

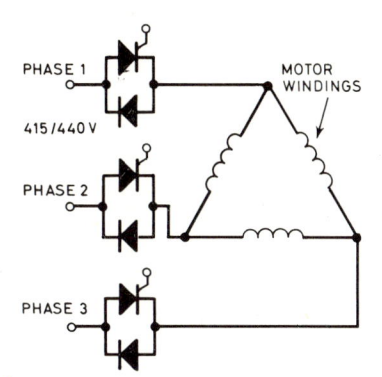

Figure 3.4. Basic arrangement of thyristors and diodes forming solid-state contactor for switching star- or delta-connected motor windings

windings. A diode is connected across each thyristor in the inverse direction to provide a return circuit for the current flow in the direction opposite to that through the thyristor. If a neutral connection is available the scheme shown in *Figure* 3.5 may be used. The particular arrangement of thyristors for a given mains voltage determines the thyristor voltage rating required.

Although the thyristor contactor has the advantage of being a high-speed switching device without moving parts, it is complicated by the ancillary components and circuits required to fire (turn on) the thyristors and protect them against adverse current and voltage conditions. For reversing duties the equipment also includes some form of logic circuitry (see page 68) to

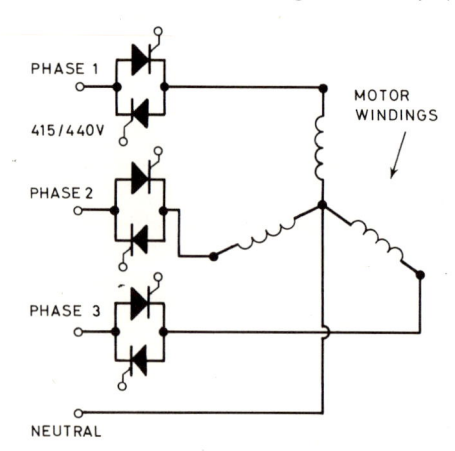

Figure 3.5. Basic arrangement of thyristors for use as contactor for switching motor windings when neutral is available

provide safety interlocking, equivalent to the tie bar which prevents mechanical forward and reverse contactors closing together in particular circumstances. However, as the equipment is entirely solid state, it is highly reliable and requires minimal maintenance, so that the relatively high cost is justified for many applications.

Thyristor Power Regulators

In addition to its switching function, the thyristor can also be used as a power regulator by varying the period of time for which it conducts during each half-cycle, so that, by controlling the instant of application of the firing pulses, the current input to the motor can be regulated from zero to a maximum. This facility provides for the 'soft' starting of motors that may not be switched direct-on-line, for controlled acceleration and deceleration, and for the speed regulation of d.c. motors supplied from thyristor converters (see page 123). Thyristors are also used in variable-frequency equipments for supplying variable-speed induction motors (see page 28).

Power regulation by varying the instant of application of the firing pulse relative to the start of the half-cycle voltage wave is referred to as phase-shift control. The time interval between the start of the half-cycle and the application of the firing pulse determines what is called the firing angle, i.e. the angle through which the supply-voltage vector has moved before the thyristor is fired. For example, with a sinusoidal voltage waveform, if the thyristor is fired as the voltage reaches its peak value, the firing angle is 90°. Theoretically, with phase-shift control the firing angle is infinitely variable from less than 1° to nearly 180°. The phase-shift control technique used for thyristor power regulators is similar in principle to that developed for the earlier valve-type power regulators.

An alternative method of power regulation uses a controller that can be adjusted to vary the number of complete half-cycles in a given period during which the thyristor conducts. With this method the supply is switched to give bursts of power of varying duration. As the thyristor can switch on and off at the same rate as the supply frequency, i.e. 50 Hz, it can be made to conduct for any number of half-cycles from zero up to 25 during a period of 0·5 s. With two thyristors in a single-phase circuit, below 100% power the output is in bursts of complete cycles the number of which is varied according to the output required.

Thyristor 3-phase Motor Switching Scheme

One example of the application of thyristor contactors is shown in *Figure* 3.6 which is a schematic diagram of the basic components and circuits of the *Ascron* static control system originally developed by Laurence, Scott & Electromotors for controlling the servomotor operating the induction regulator used with the N-S variable-speed 3-phase commutator motor (see page 37). The *Ascron* system is also used for other induction-regulator applications, e.g. voltage or power-factor control, and for motor-driven actuators for modulating duties.

With the scheme shown in *Figure* 3.6, a 3-phase squirrel-cage servomotor is switched by an arrangement of thyristors controlled by transistorised circuits responding to a low-level signal input. One phase of the servomotor is connected directly to the supply, the other two phases are connected via four contactors each comprising two thyristors connected in inverse parallel. When operating, each pair of thyristors will maintain an a.c. supply to the servomotor.

For rotating the servomotor in one direction, contactors 1 and 2 are operated by firing pulses from one generator. To rotate the servomotor in the opposite direction, its phase sequence is reversed by operating contactors 3 and 4 by firing pulses from the other generator. With this arrangement the servomotor rotates in either direction according to which pulse generator receives a d.c. input signal from the amplifier.

The firing signals are applied to the thyristor gates as a continuous train of high-frequency pulses to ensure that each thyristor fires early in the appropriate half-cycle. The input to the servomotor is in the form of bursts of complete cycles of power, and regulation is effected by on/off switching to vary the duration of the burst according to the control action demanded by the error signal. With this regulating technique the full voltage is applied to the servomotor so that it always develops full torque no matter for how short a period the supply is switched on. Also, when the control system is in balance the servomotor is completely de-energised.

With the phase-shift technique, power is regulated by varying the mean value of the voltage during each half-cycle so that for minimum power output the voltage is also a minimum. Moreover, as the power is reduced

26   A.C. *Motors and Power Regulators*

there is a phase shift of the current with respect to the applied alternating voltage, consequently the power factor also decreases.

Operation of the pulse generators of the *Ascron* system is controlled by output signals from the amplifier resulting from an input signal which is the error voltage obtained by comparing the demand-signal and reset-signal

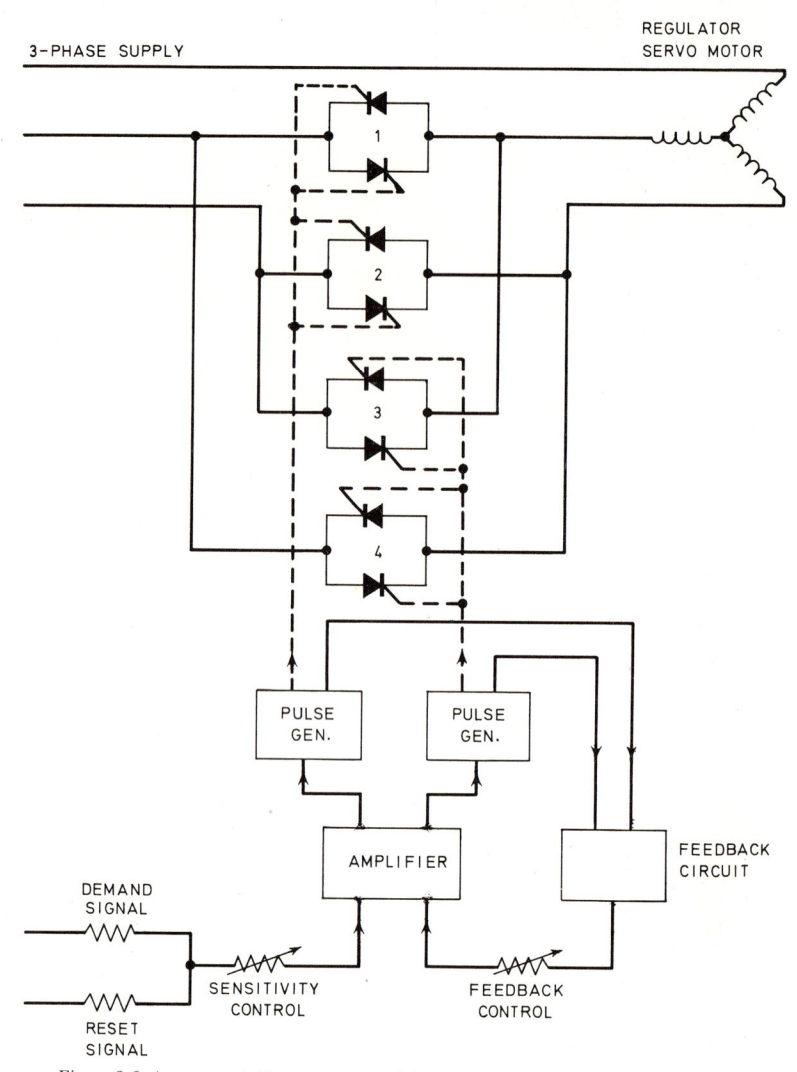

Figure 3.6. Ascron *switching system comprising thyristors controlled by pulse generators*

voltages. These voltages are proportional to the desired and measured values of the controlled quantity respectively. The reset signal is derived from devices sensing and measuring changes in the controlled quantity.

From each pulse generator there is a stabilising signal feedback into the amplifier which causes the servomotor to be switched with a ratio of on-time to off-time which depends on the magnitude of the error signal. For large errors the on-time is relatively long but as the error diminishes the ratio decreases so that the final adjustment of the servomotor is made in smaller and smaller increments to ensure that it is positioned precisely and rapidly. For small errors the system has a fully proportional control action without any tendency to over-correction of the error signal which would cause the servomotor to rotate forward and backward so that the system would be unstable.

Squirrel-cage Motor Reversing Drives

For many simple reversing drives the control requirement is met by using a squirrel-cage motor and a reversing direct-on-line starter. To achieve rapid reversal or stopping, braking can be applied either by the plugging or d.c. injection methods.

Plugging involves switching the winding connections for reverse rotation while the motor is running forward. If the motor only has to be stopped rapidly, to prevent it reversing from standstill a shaft-mounted centrifugal switch ensures that the windings are disconnected at the appropriate moment.

With d.c. injection braking, the motor stator is disconnected from the supply and reconnected to a d.c. supply derived from a rectifier unit which is switched in as the main contactor opens. As the stator becomes a d.c. field system the rotor functions as a generator armature driven by the load and therefore subjected to a powerful braking torque. In general, the method provides more precise, rapid and smoother braking than plugging without the risk of the motor reversing.

In the case of pole-changing motors (see below) there is an automatic braking action when changing to a lower speed because as the reconnection of the windings produces a larger number of poles, initially the rotor is rotating faster than the corresponding field and consequently it functions as a generator armature.

Multi-speed Squirrel-cage Motors

Apart from being used for on/off single-speed drives, squirrel-cage motors are designed to provide a choice of two, three or four speeds. In an induction motor the number of pairs of stator poles determines the synchronous speed, and hence the speed can be altered by changing the number of pole pairs. This is achieved by special stator winding arrangements with connections to speed selection switches.

One form of winding gives two different speeds in the ratio 2 : 1. It may be wound, for example, for two and four poles so that the synchronous speeds at 50 Hz are 3000 and 1500 rev/min—the actual speeds being slightly less due to the slip essential to the development of torque. With four and eight poles, six and twelve poles, and so on, the two corresponding speeds are obtained.

Motors with two superimposed windings are designed to operate at two widely different speeds, e.g. with 2 and 24 poles at synchronous speeds of

3000 and 250 rev/min, at 50 Hz. With another arrangement of the two windings three or four different speeds are obtained.

Developed as an alternative to the two-winding motor, the PAM (pole-amplitude modulated) motor provides two fixed speeds at ratios other than 2 : 1 from a single tapped winding. The first PAM motors were designed for two close-ratio speeds, i.e. those obtained with four and six poles, six and eight poles, eight and ten poles and so on. These combinations are used extensively, and additionally wide-ratio combinations are obtainable in all ratios except 3 : 1, and three speeds can be provided with the single winding.

Variable-frequency Induction-motor Drives

As with a given number of poles the speed of an induction motor depends on the supply frequency, it follows that when operated from a variable-frequency supply the motor can be used for drives with a variable-speed requirement that can be met economically employing equipment powered from the standard frequency mains and producing the necessary frequency range.

For some drives the speed required is such that the frequency range is below the standard frequency while for others it is above. At the same time the requirement may be for a constant speed at either the lower or higher non-standard frequency.

Although the use of a variable-frequency supply would enable the simple squirrel-cage motor to be applied to many variable-speed drives, it is a practical proposition only when the speed of a large number of small motors has to be regulated simultaneously; and perhaps when the size of a single drive motor has to be minimised, or the motor is operated in an environment where sliding contacts are not permissible. Except in these special circumstances, the cost of the variable-frequency supply equipment does not generally justify its use for installations requiring only a few squirrel-cage motor drives.

Variable-frequency Supply Equipments

The variable-frequency supply equipments used for specific applications are shown schematically in *Figure* 3.7. They are also used for non-standard constant-frequency supplies.

A conventional alternator generates at a frequency proportional to its rotational speed so it provides a variable-frequency supply when driven by a variable-speed motor. This may be a 3-phase commutator motor or a rectifier-fed d.c. motor. *Figure* 3.7(a) shows a scheme comprising a 3-phase commutator motor, with an induction regulator for speed variation, driving a compounded a.c. generator. This type of generator has a voltage/frequency characteristic especially suitable for the efficient operation of low-frequency motors. Compounding is effected by an additional series-connected field

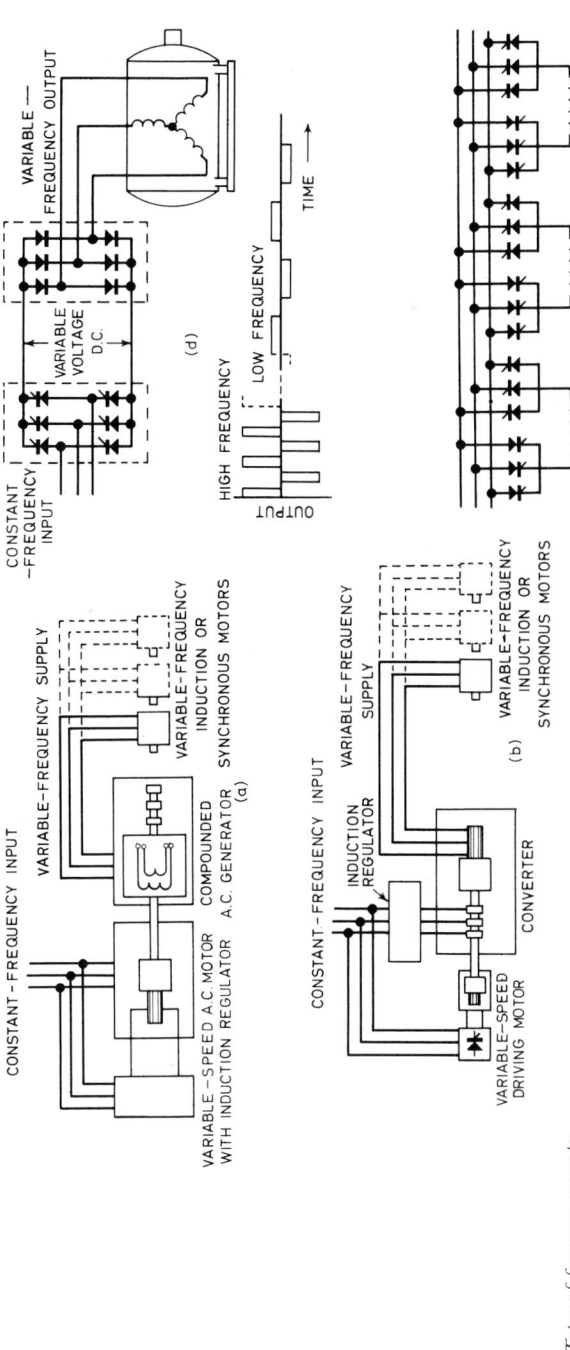

Figure 3.7. Types of frequency converter

(a) motor–generator set; (b) pony motor driven commutator machine consisting of an armature rotating within an unwound stator; (c) self-driven converter similar to Figure 3.7 (b) but with a stator winding in the converter; (d) static d.c. link converter; the output frequency is changed by chopping the variable voltage direct current; (e) cyclo-converter, which produces a low-frequency output from a 50 Hz input.

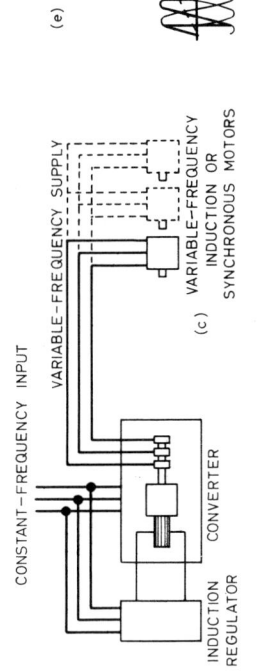

winding which is fed (by means of a current transformer and rectifier) with d.c. proportional to a.c. load.

Rotating Frequency Converters

Figure 3.7(b) shows the main components and connections of the N-S frequency converter designed and manufactured by Laurence, Scott & Electromotors Ltd. The converter consists of a special rotor with a commutator and sliprings; a stator which has only a starting winding; a small variable-speed driving motor; and an induction regulator supplying a 3-phase variable-voltage to the converter sliprings.

The 3-phase supply to the rotor windings produces a magnetic flux which rotates at synchronous speed N_1 relative to the rotor core, and at the same speed relative to the stator core when the rotor is stationary. If then the rotor is driven at a speed N_2 in the opposite direction to the rotation of its own flux, the magnetic field crossing the air gap and passing through the stator core will rotate at a speed $N_1 - N_2$. When the rotor speed N_2 approaches the synchronous speed N_1, a low-frequency flux of $(N_1 - N_2)/N_1 \times 50$ Hz will be obtained in the stator core. The effect of this flux is to produce a low-frequency output at the brushes by the action of the commutator.

Under normal running conditions it is only the core of the stator that is utilised for completing the flux path—the winding is in circuit only during starting. Frequency conversion is therefore effected electromagnetically so that the torque developed by the driving motor is only that required to overcome the mechanical resistances to rotation. Consequently, the converter driving-motor rating is much less than that required to drive an alternator with the same output.

As the output frequency falls as the rotor speed increases from zero and approaches the synchronous speed, it can be varied by adjusting the driving-motor speed. The converter output voltage is varied by altering the 50 Hz input voltage to the sliprings.

With another version of the converter the supply is fed to the sliprings through an input transformer instead of an induction regulator. This arrangement serves to provide for a supply at the appropriate voltage but does not allow for the variation of the voltage to obtain a constant voltage/frequency characteristic which is necessary for some drives. The advantage of the induction regulator scheme is that any voltage/frequency characteristic can be obtained.

In principle, this type of converter can operate over a wide frequency range, but in practice it is especially suitable for ranges of about 2·5–25 Hz for which motor–alternator sets are more expensive.

Figure 3.7(c) shows a type of frequency converter which has a stator winding, so that it operates as a variable-speed a.c. commutator motor with induction-regulator control instead of being driven by a pony motor. The only mechanical power developed by the converter is that required to drive the rotor to give a variable-frequency output at the sliprings. The induction regulator varies the speed and therefore the output frequency. This type of converter is used for the range 5–45 Hz and has an inherent constant voltage/frequency characteristic.

Static Frequency Converters

Static equipments incorporating thyristors are also being used for particular applications where they have certain advantages compared with rotating converters.

Figure 3.7(d) shows a d.c. link type of static converter used for ranges of, say, 10–1000 Hz or more. It consists of a 3-phase bridge arrangement of thyristors which converts the constant-frequency input to a variable-voltage d.c. supply. This is fed to a second thyristor bridge, the firing sequence of which is controlled in such a way that its output consists of a series of substantially square-shaped pulses whose frequency is varied to provide the supply to the induction motor.

Simple d.c. link converters of this type cannot operate regeneratively, i.e. they do not allow power to be fed back into the supply system when the motor is functioning as a generator, e.g. when it is being stopped by dynamic braking. It is an advantage of rotating converters that they can accept regenerated power. If this facility is required with a d.c. link converter system two identical equipments must be connected in inverse parallel to allow for the flow of power in both directions. This arrangement is also used with a.c./d.c. rectifiers to allow for braking and reversing of d.c. motors (see page 59).

Because the output waveform of the d.c. link converter is substantially square, induction-motor performance becomes increasingly impaired as the supply frequency is reduced. This sets the lower frequency limit of d.c. link converters, but for the lowest frequency range, say 1–25 Hz, the static cycloconverter is available.

Shown in *Figure* 3.7(e), the cycloconverter consists of two complete thyristor bridges for each phase of the supply. The firing of the individual thyristors of the six bridges is controlled so that the output of the converter consists of a low-frequency wave built up from small portions of the individual input waves. Although the large number of thyristors required makes the cycloconverter relatively expensive, it has the advantage of accepting regenerative power.

Variable-frequency Roller-table Motors

In general, a variable-frequency scheme becomes feasible when variable-speed induction motors are the best type for a particular drive and the cost of the equipment can be shared between a sufficient number of motors. A typical application is to roller-table drives in steelworks where the variable-frequency supply for simultaneous speed regulation of a large group of motors is a much better proposition than using a speed-control scheme for each motor.

Figure 3.8 shows the characteristic curves for two different frequencies of roller table motors produced by Laurence, Scott & Electromotors Ltd.

Variable-frequency Synchronous-motor Drives

A variable-frequency supply provides for the simultaneous speed regulation of small synchronous motors used to maintain several drives at precisely the same speed. This requirement cannot be met with induction motors, owing to slight differences in slip.

For this application the 3-phase synchronous reluctance motor has the advantage that having an induction-motor type stator it can be started

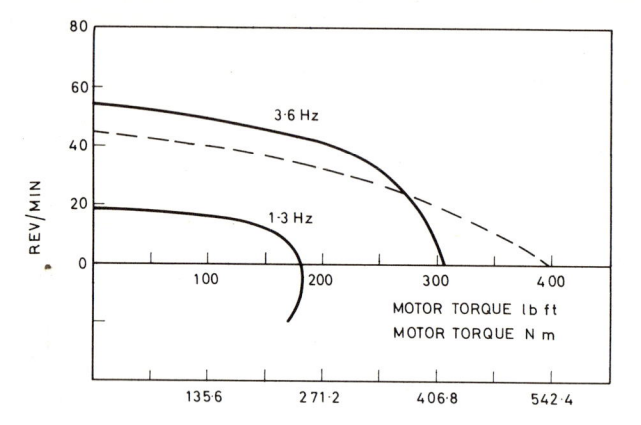

Figure 3.8. Roller-motor speed/torque characteristics

direct-on-line although it runs as a true synchronous motor at a speed precisely related to the supply frequency.

Typical synchronous reluctance motors are those in the Brook Motors *Relsyn* range available with rated outputs from f.h.p. up to 15 kW (20 hp) with 50 Hz speeds of 3000, 1500 and 1000 rev/min. This type of motor can also be produced in ratings up to 56 kW (75 hp) or more. *Figure* 3.9 shows the torque/speed characteristic. Pull-out torque is more than 150% full-load torque, i.e. the motor can develop up to at least 1·5 times rated torque before it pulls out of exact synchronism.

Operated from thyristor frequency converters with a square-wave output, *Relsyn* motors perform well down to about 10 Hz although the losses increase. On sinewave supplies lower frequencies can be used. The motors can be started at low frequency to limit the starting current.

Torque Motor Drives

Squirrel-cage torque motors of special design used for reeler drives requiring up to about 2·24 kW (3 hp) have an inherent speed regulation, which enables constant tension to be maintained over a coil build-up range of 2 : 1.

A.C. Motors and Power Regulators

Figure 3.10 shows the speed/torque characteristic of an English Electric torque motor for this application. The characteristic can be changed by variation of the applied voltage so that with a suitable voltage regulator the motor can be used for drives where a range of tensions and speed of material feed are required.

A closed-loop controller developed by Failsafe Automation for use with

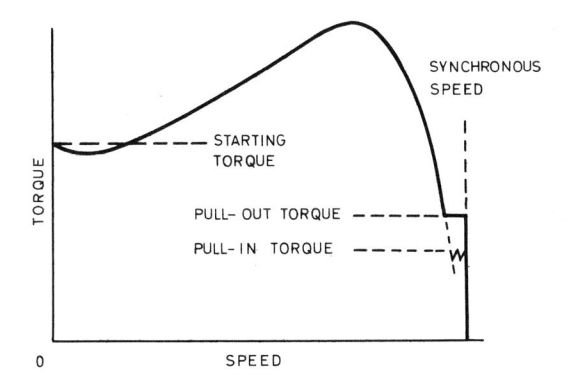

Figure 3.9. Relsyn-motor torque/speed characteristic

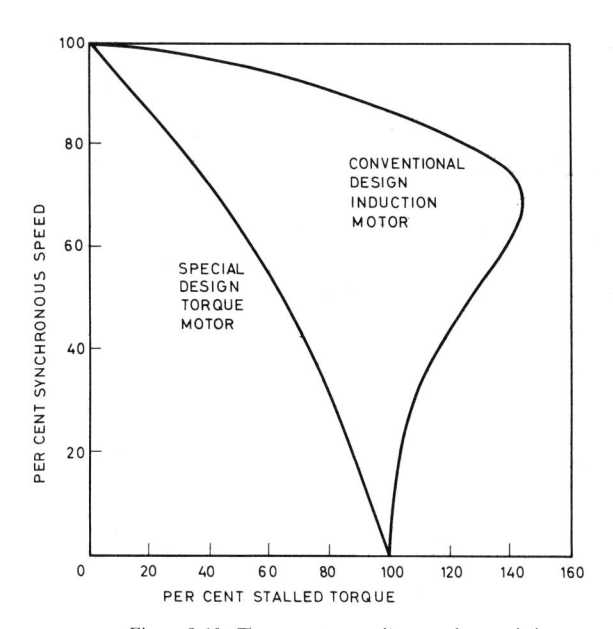

Figure 3.10. Torque-motor speed/torque characteristic

the English Electric torque motor provides for voltage regulation within ±5 V at any set point within the range. With a tachogenerator feedback from the motor, a speed regulation of ±1% is obtained over a 5:1 speed range with load changes of from 25 to 100% load. The controller can also be used to control the motor current to regulate the motor torque.

The torque motor is also used for applications where it is required simply to apply torque in a stalled condition against an opposing torque so that it does not rotate unless the input current changes. An application of this type is shown in *Figure* 3.11.

Slipring Induction-motor Drives

The control characteristics of the slipring induction motor are similar to those of the squirrel-cage motor but they can be modified to some extent by adjustment of the external rotor resistor used to reduce the starting current and, if required, to provide a limited range of speed regulation. The resistor may be a metallic or a liquid type, the latter being used to give stepless speed regulation.

The slipring induction motor is used to drive the generator of a Ward-Leonard set, and also in the similar Ilgner set which includes a flywheel providing energy during the peak loading period of the duty cycle to reduce the power input to the motor. To enable the flywheel stored energy to be released, the slip of the motor must increase automatically with the load. This requirement is met with the scheme shown in *Figure* 3.11 by the torque motor adjusting the liquid resistor slip regulator in the rotor circuit as the stator current increases. The torque motor is energised from current transformers. The pilot motor drives a winch which can be used to adjust the liquid resistor for starting-current limitation or controlled dynamic braking of the set.

Variable-speed Couplings

A combination of a squirrel-cage motor and an electrical, mechanical or hydraulic type of variable-speed coupling is used for particular industrial drives. In effect, the coupling is a power regulator as it regulates the power transmitted to the driven unit and therefore the input to the motor.

For automatic drives the electrical eddy-current type of coupling is used since it can be controlled readily by signals from an electrical/electronic scheme. *Figure* 3.12 is a schematic diagram relating to the operation and control of an eddy-current type coupling produced by Mawdsley's Ltd. in ratings up to 450 kW (600 hp) or more.

With a constant-speed induction or synchronous motor drive, the output speed is regulated by variation of the excitation current input to the stationary field winding (1). The excitation current is supplied from a d.c. source. A rotor (2) is fitted to the output shaft. Relative motion between the motor-driven inductor drum (3) and the rotor causes a small section of the drum to rotate past alternate north and south magnetic poles. Eddy currents induced in the drum by the relative motion interact with the field and create a force enabling torque to be transmitted between the input and output shafts. The torque transmitted is proportional to the excitation current so this is varied to regulate the torque.

With a constant-torque load, speed regulation is effected by increasing the excitation to provide the excess torque for accelerating the driven unit to a

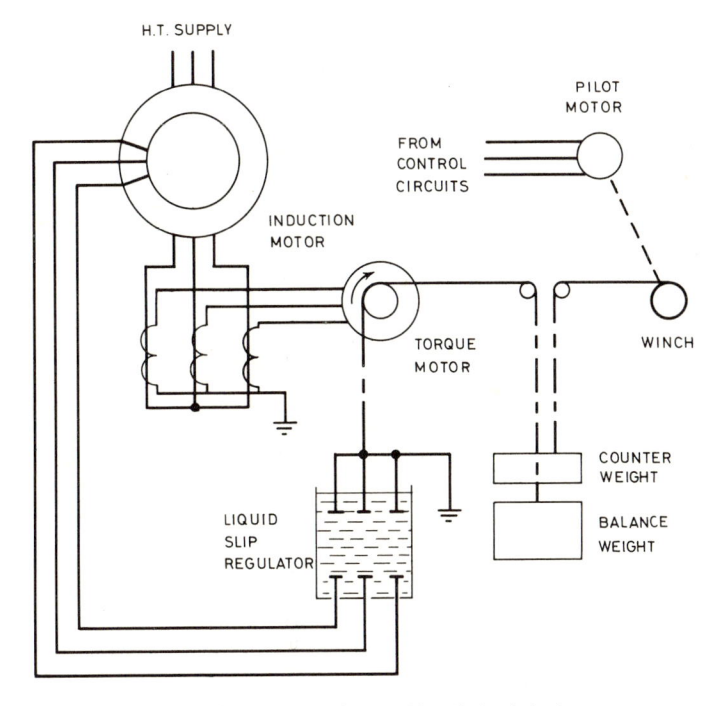

Figure 3.11. Slip regulator of Ilgner drive slipring induction motor

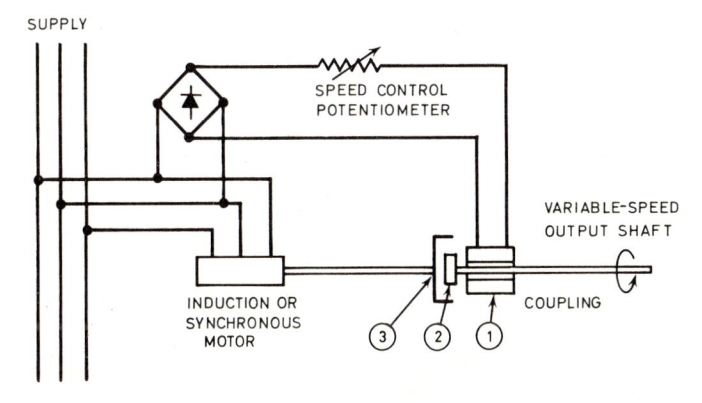

Figure 3.12. Schematic diagram of eddy-current coupling and control system

higher speed, or decreasing the excitation to reduce the torque so that the driven unit decelerates. As the higher or lower speed is reached the excitation is adjusted to the value necessary to maintain the load torque.

As the generation of eddy currents depends on relative motion between the rotor (2) and the inductor drum (3), the coupling has an inherent slip which varies with the torque transmitted. Increasing load on the output shaft increases the slip and therefore the magnitude of the eddy currents generated and the torque transmitted. To maintain a given speed the excitation must be increased. When the coupling is driven by an induction motor its inherent slip also affects the speed/torque characteristic of the drive.

In *Figure* 3.13 curve A is the torque/speed characteristic of the motor and curve B is that of the coupling at full excitation. Curves C and D are the

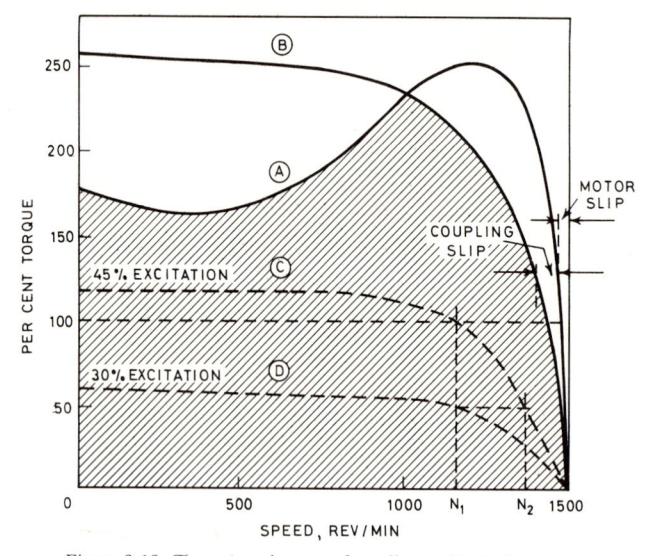

Figure 3.13. *Torque/speed curves of coupling and induction motor*

torque/speed characteristics of the coupling for 45% and 30% excitation respectively. The diagram also shows the variation of motor and coupling slips with torque and speed.

Figure 3.13 shows that for an output shaft speed N_1 an excitation of 30% is required to transmit 50% of full-load torque. To obtain 100% torque and maintain the speed at N_1 the excitation is increased to 45%. If then the torque output reduces to 50% and the excitation is held at 45%, the output shaft speed rises to the value N_2.

With an automatic control scheme, speed and torque can be regulated automatically by varying the excitation of the coupling by the speed-control potentiometer being adjusted in accordance with a feedback signal. For speed regulation a scheme including a tachogenerator and an amplifier controls the excitation. Output torque can be regulated in terms of the motor input current which can be measured by some device providing a suitable feedback signal for the control scheme. This signal can also be used to provide overriding current-limit protection for a speed-control scheme.

As the heat resulting from the eddy currents has to be dissipated, the couplings must be designed to provide for effective air or water cooling. Couplings in Mawdsley's range are air-cooled up to about 75 kW (100 hp) and water-cooled for up to 450 kW (600 hp) or more. The output speed range is about 1410–80 rev/min with a 4-pole driving motor running at 1460 rev/min. Using an a.c. type tachogenerator standard regulation is $\pm 2\%$ for 75% steady-state load changes but $\pm 0.5\%$ and $\pm 0.1\%$ can be obtained if required.

A particular facility of the coupling is that it provides for the slow smooth starting of the driven unit. For example, it is used for variable-speed textile machines for which other drives had too high an initial starting torque that resulted in snatching and/or breaking of the yarn at one or more positions on the machine. For this class of application the motor is started and the coupling output is accelerated smoothly over the period of time necessary to avoid damage to the product.

The eddy-current coupling has no inherent braking facility, but this can be provided by an eddy-current brake fitted to the output shaft of the coupling. The brake unit is similar in principle and construction to the coupling comprising a rotating drum and stator with exciting winding. Braking is achieved by the interaction of eddy currents in the drum with the stationary field system.

With one type of brake produced by Mawdsley's only the drive coupling is energised while the machine is being driven, the brake field excitation being switched off. To stop the drive the coupling excitation is switched off and the brake excitation switched on. No control can be exercised by a closed-loop scheme during deceleration, the braking torque simply reduces exponentially with the speed. The brake meets the requirement for rapid but uncontrolled braking, e.g. for emergency stopping.

With a second type of Mawdsley's brake unit closed-loop, the brake field excitation is controlled by a closed-loop scheme to ensure smooth and precise deceleration.

Variable-speed 3-phase Commutator Motors

For a variety of drives variable-speed 3-phase commutator motors are used either to maintain a constant speed at any value in the speed range, or to vary the speed of the driven unit in accordance with changes in the desired value of the controlled quantity.

In principle, the 3-phase commutator motor is equivalent to a slipring induction motor with its rotor winding connected to a variable-voltage source which is used to buck or boost the induced rotor e.m.f. and thereby lower or raise the rotor speed below or above the synchronous value. The method of varying the rotor voltage of the 3-phase commutator motor depends on whether it is a stator-fed or a Schrage type.

Stator-fed 3-phase Commutator Motors

With the stator-fed type the speed-regulating variable-voltage is obtained from the secondary-winding of either an induction regulator or a variable-ratio transformer. The regulator gives stepless regulation while with the transformer regulation is in increments, the value of which depends on the number of tappings provided. The stator of the motor and the primary winding of the induction regulator or the transformer are connected to the 3-phase supply and can be wound for up to about 11 kV.

Figure 3.14 shows schematically the main circuits of a stator-fed motor with its rotor supplied from an induction regulator. A 3-phase voltage at supply frequency is fed to the rotor winding and the induced e.m.f. appearing at the brushes is also at supply frequency because the commutator

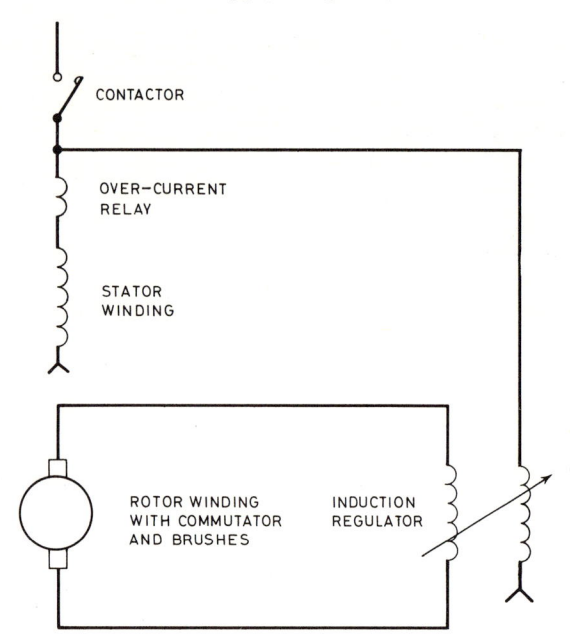

Figure 3.14. Basic connections of stator-fed 3-phase commutator motor

functions as a frequency converter. The induced e.m.f. results from the difference between the speeds of the rotating stator field and the rotor so that its frequency varies with the rotor speed, and conversion to the normal supply frequency is necessary to enable the applied voltage to be used to regulate the speed and power of the motor.

At one time, owing to the difficulty of obtaining an adequate standard of commutation, 3-phase commutator motors had to be designed with a comparatively low output per pole. With this restriction large outputs could be achieved only by increasing the number of poles with a consequent limitation of the speed range.

With the N–S type stator-fed motors of Laurence, Scott & Electromotors, the limitation to the permissible output per pole has been overcome to a very large extent by the use of an auxiliary commutator winding in the rotor of the machine. This winding is located at the bottom of the rotor slots and is connected to the same commutator segments as the main rotor winding. The function of the winding is to absorb the energy of commutation and deliver it by transformer action to the groups of the main winding conductors in the slots not at that moment undergoing commutation. The auxiliary winding also deals with the secondary but important influences of parasitic fluxes on the commutation of the motor by damping out the higher harmonic content of the main flux.

N–S motors are available in designs for outputs up to 3730 kW (5000 hp) at top speeds approaching 600 rev/min and speed ranges of 10 : 1 or more, depending on the load characteristic. The top range can be extended up to about 7500 kW (10 000 hp) by the use of double armature or tandem equipments. The stator winding can be wound for 3·3 kV, 6·6 kV and 11 kV, at

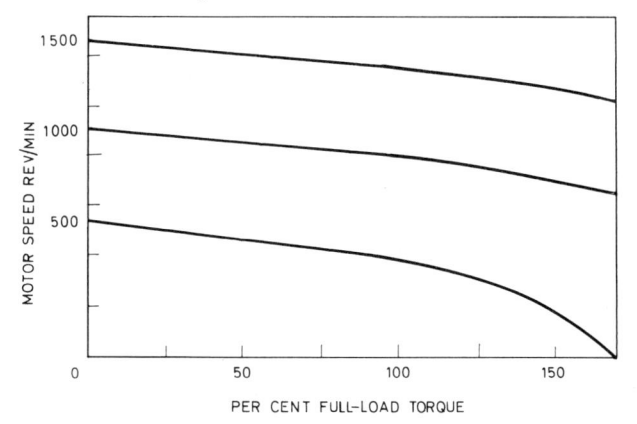

Figure 3.15. Torque/speed characteristics for 1440/480 rev/min N-S motor at various regulator positions

which voltages the minimum economic outputs are similar to those for induction motors. Starting is normally direct-on-line with a starting current in the order of 1½–2½ times full-load current.

The normal N–S motor has a shunt speed characteristic, i.e. the speed falls slightly with increasing load. The shape of the inherent torque/speed characteristic varies only slightly with different regulator adjustments. *Figure* 3.15 shows typical characteristics. The N–S motor can also be designed with series characteristics when required.

For most of the industrial drives for which the N–S motor is suitable, a shunt characteristic is required so that at any given regulator setting the speed remains substantially constant irrespective of variations in load. Such characteristics are obtained inherently with a normal N–S motor and induction regulator connected as shown in *Figure* 3.16.

Induction Regulators

An induction regulator is basically a transformer but is similar in construction to a 3-phase induction motor with primary and secondary windings, the primary usually being on the rotor and the secondary on the stator. The rotor is moved relative to the stator to vary the output voltage of the secondary winding.

The double induction regulator (*Figure* 3.16) consists of two conventional phase-shifting single induction regulators built as a single unit with a common shaft. When the rotor of a single induction regulator is displaced from the neutral position, the magnitude of the secondary voltage is not changed but its phase is rotated by the angle of displacement. With two such regulators mounted on one shaft with their windings suitably connected the variation in phase position of the individual secondary voltages can be arranged to vary the output voltage from a maximum in one direction through zero to a maximum in the opposite direction. The phase of the resultant output

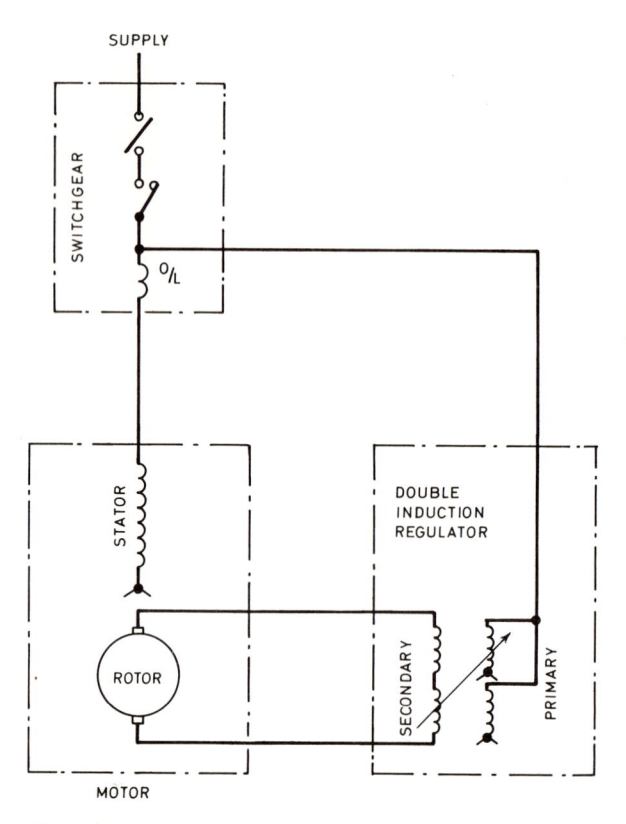

Figure 3.16. N-S motor with double induction regulator (one phase only shown)

voltage remains constant irrespective of the magnitude of the output voltage, an important feature of any regulator used for speed regulation of a 3-phase commutator motor since, unless the phase position remains substantially constant, it becomes difficult to control the power factor and current magnitudes in the equipment.

Because the cost of the double induction regulator is relatively high, single regulator schemes were developed to provide for efficient speed regulation of N–S motors. With the scheme shown in *Figure* 3.17 the regulator stator and rotor windings, connected in series, are fed from an auxiliary winding located in the stator of the N–S motor. The output voltage appears between the centre tapping A on the motor auxiliary winding and the junction B between the regulator stator and rotor windings. When the rotor is displaced from the neutral position, the regulator stator and rotor voltages will be at an angle corresponding to this displacement, consequently a voltage will appear between the points A and B which, when applied to the motor rotor, will vary the motor speed.

When the single regulator scheme is used for motors above, say, 75 kW (100 hp), it becomes more convenient to feed the induction regulator from a separate transformer instead of from the auxiliary winding in the motor

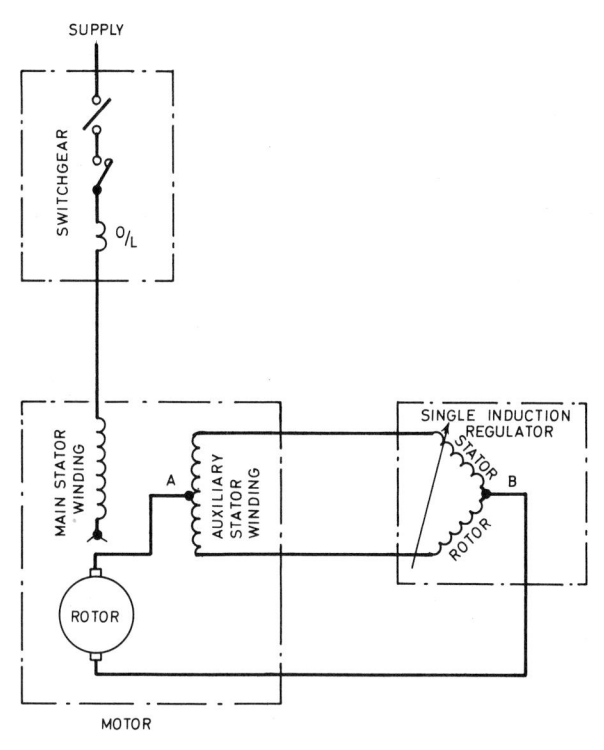

Figure 3.17. N-S motor with a single induction regulator fed from a motor stator auxiliary winding

stator. With this arrangement (*Figure* 3.18) the supply to the regulator can be at a voltage lower than that of the mains supply to the motor stator which is an advantage when the mains supply is at a relatively high voltage. The regulator is therefore always wound at a voltage level corresponding to that of the motor rotor circuit and can be constructed more compactly and conveniently than if it were insulated for the full supply voltage. In schemes using double induction regulators the primary must always be wound for the supply voltage so that occasionally the regulator must be larger than is really necessary in order to accommodate the high voltage. Alternatively, it may be necessary to use a step-down transformer to feed the regulator.

Normally a compensating voltage is injected into the motor rotor winding to improve the overall power factor of the equipment and to reduce the on-load rotor current, thereby improving the torque characteristics of the motor. The compensating voltage may be obtained from a compensating winding in the motor stator or from a small compensating transformer. A double induction regulator can be designed to give an output voltage with a power-factor correcting component.

With single induction regulator schemes power-factor correction voltages are obtained by displacement of the tapping point away from the centre of the winding supplying the regulator. With this arrangement any required degree of power-factor correction can be obtained without increasing the cost of the equipment.

When the speed of several driven units has to be varied simultaneously a group of N–S motors can be fed from a common regulator. If required

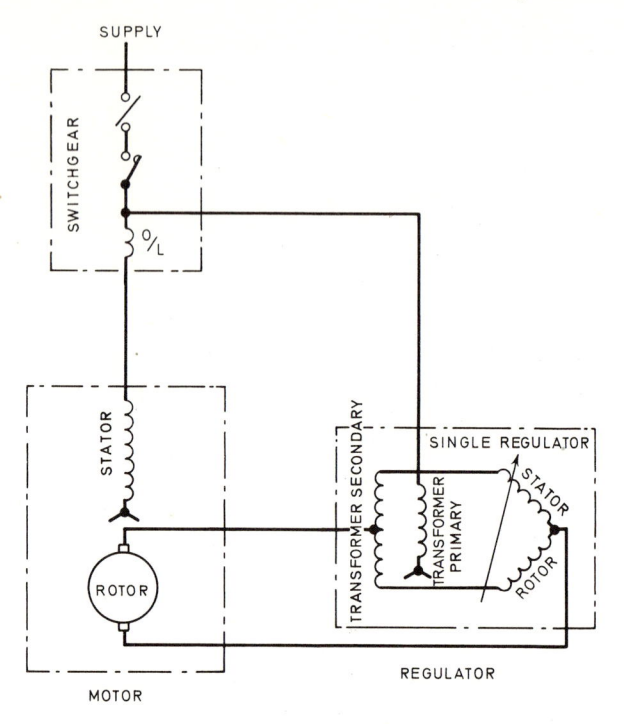

Figure 3.18. N-S motor with single induction regulator fed from a transformer in the regulator bedplate

trimming regulators can be used to enable the speeds of the individual motors to be trimmed above or below the basic speeds set by the common regulator.

If a continuous speed range is not required, for some drives the induction regulator can be replaced by a tapped transformer, or by a tapped auxiliary winding in the motor. Using rotary switches any one of a number of fixed speeds may be selected. Speed changes are made rapidly and with contactor tap-changing high values of acceleration and deceleration are achieved.

For most drives powered by N–S motors sufficient starting torque is developed by direct-on-line switching with the induction regulator in the lowest speed position, as shown by the bottom curve in *Figure* 3.15. To ensure low-speed starting, an electrical interlock with the starter can be fitted to the induction regulator. Other forms of starting equipment are available to maximise starting performance—either to obtain high starting torques, or to reduce the starting current.

In conjunction with suitable switchgear, N–S motors may be designed for certain reversing duties.

Regenerative braking is inherent in the design of the N–S motor and for most applications normal deceleration is controlled adequately. Controlled rapid retardation may be obtained by the d.c. injection dynamic braking method.

Speed Regulation of N–S Motors

The speed of the N–S motor may be regulated automatically by control of the pilot motor driving the induction regulator rotor, or by a slave cylinder actuated by hydraulic or pneumatic pressure and providing direct mechanical movement of the induction regulator rotor.

For the automatic control of N–S motors Laurence, Scott & Electromotors has developed special schemes which control the operation of the pilot motor driving the induction regulator rotor. The schemes provide for the closed-loop control of any quantity that can be maintained at a desired value by regulating the speed of the N–S motor.

The L.S.E. *Ascron* control scheme (see page 25) has thyristor contactors for switching the pilot (servo) motor, and there is also the *Asrec* scheme (see page 93) using standard polarised relays.

Motor speed can also be regulated by an open-loop scheme, which enables a desired speed to be selected before starting or reset during operation by moving the pointer of a calibrated scale. Local and remote pushbutton controls provide for starting, accelerating, retarding and stopping the motor.

As the induction regulator is also used to provide a variable-voltage supply for such purposes as the regulation of the input to industrial electrical heating equipment, the combination of the pilot motor and its control scheme is in effect an automatic drive. Similarly, the combination of the control devices and servo motor of the *Ascron* scheme, for example, is also in effect an automatic drive which is used in the form of an actuator for mechanical valves and dampers.

Schrage-type 3-phase Commutator Motors

Compared with the stator-fed type of 3-phase commutator motor, the Schrage type has the advantage that it does not require a separate power regulator, but since the supply input is to the rotor through sliprings it can be wound only for up to about 650 V whereas the stator-fed type can be supplied at up to about 11 kV.

Typical of the Schrage motors used for a variety of automatic industrial drives, the type CH motors are available from English Electric–AEI Machines in ratings up to 373 kW (500 hp). The type CH motor has an arrangement of windings shown diagrammatically in *Figure* 3.19. The primary winding on the rotor is equivalent to the stator of an induction motor. The rotor also has a regulating winding, connected to the commutator, providing an e.m.f. which is injected into the secondary winding through the brushes.

Since the 3-phase supply to the rotor produces a field rotating relatively to the secondary winding (which is equivalent to the rotor winding of an

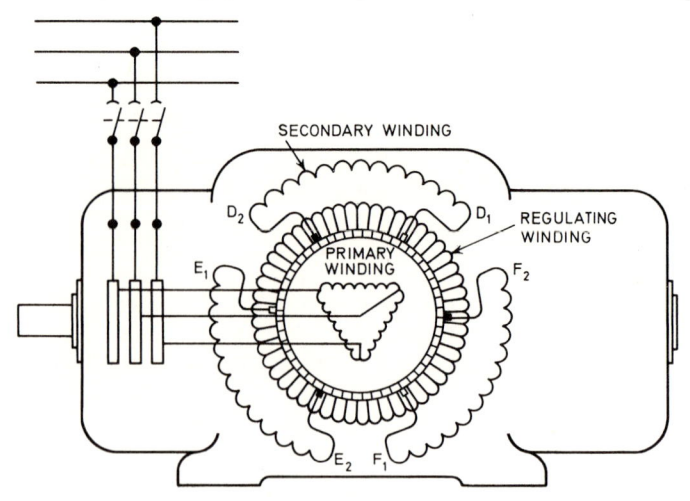

Figure 3.19. Windings of type CH 3-phase commutator motor

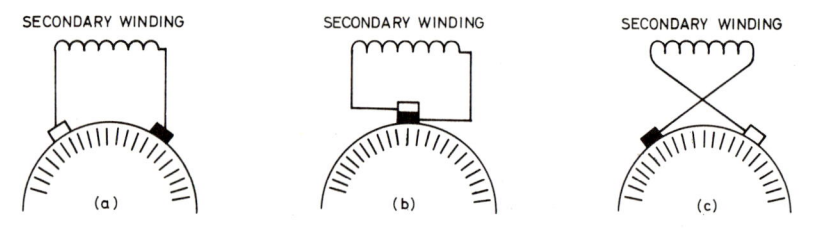

Figure 3.20. Brushgear positions of type CH motor for super-synchronous, synchronous and sub-synchronous speeds
(a) Positive injected voltage above synchronous speed; (b) Zero injected voltage synchronous speed (less slip); (c) Negative injected voltage below synchronous speed

induction motor), an e.m.f. is induced in the secondary at a slip frequency determined by the rotor speed. The commutator functions to maintain the injection e.m.f. frequency at the same value as the slip frequency.

Speed Regulation of Type CH Motors

The magnitude of the injected e.m.f. is varied by moving the two sets of brushgear relative to one another to obtain infinite variation of the rotor speed over a range above and below the synchronous speed of the motor when operating as an induction type. This occurs when the two sets of

A.C. Motors and Power Regulators

brushgear are moved around the commutator simultaneously to the position shown in *Figure* 3.20. The secondary winding is then equivalent to a squirrel-cage rotor, and as the injected e.m.f. is zero, the rotor is running at synchronous speed less slip. *Figure* 3.20 also shows the relative positions of the two sets of brushgear to obtain super-synchronous and sub-synchronous speeds.

The brushgear rockers can be moved either manually by a handwheel or by a pilot motor when speed is regulated by pushbutton control or by an automatic scheme. *Figure* 3.21 shows a typical pilot motor control scheme for a forced ventilated motor with raise/lower pushbuttons for speed adjustment. The scheme includes a low-speed interlock and limit switches which operate at each end of the brushgear travel. These are additional to mechanical stops to prevent over-travel. Another two switches can be fitted at intermediate brushgear positions when required for special control purposes;

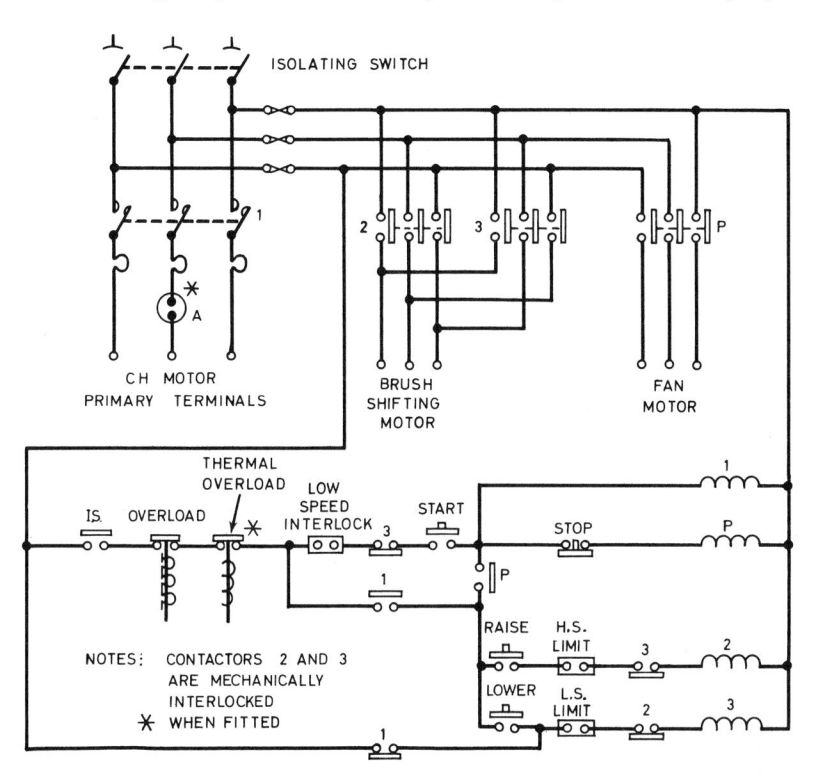

Figure 3.21. Typical control circuits of type CH motor

e.g. some applications require the motor to start above its minimum speed setting, or it may be necessary to interlock another drive so that it cannot function until the motor has reached a particular speed. Preset speed regulators are used for various automatic control schemes.

The AEI type CH motors have an inherent constant torque characteristic (i.e. the power output is directly proportional to the operating speed) and develop from 1 to $1\frac{1}{2}$ times full-load torque at starting depending upon the brushgear position. The standard method of starting is direct-on-line with the brushgear set at the minimum speed.

The motors are suitable for either direction of rotation but it is usually necessary to derate motors required for frequent reversing drives. They can be supplied for regenerative braking with suitably designed control gear. Alternatively, a disc brake can be fitted to a shaft extension at the non-drive end. Extremely rapid braking can be applied by plugging with control gear to reverse the supply to the sliprings and switch a resistor into the secondary circuit to limit the current to a safe value.

The speed range available depends on the output required at the top and bottom speeds. For example, a 2 : 1 range motor has outputs of 67 kW (90 hp) at 1440 rev/min and 33·5 kW (45 hp) at 720 rev/min; with a 20 : 1 range motor the outputs are 45 kW (60 hp) at 1700 rev/min and 2·2 kW (3 hp) at 85 rev/min.

Chapter 4

D.C. Motors and Power Regulators

For many drives d.c. motors are far more effective than any a.c. types because they enable the output of the driven units to be maximised. Although the initial cost of a d.c. motor scheme may be higher because it includes rectifier equipment, if the gain in production increases the long term profitability then the cost is justified.

In general, compared with any type of a.c. motor, the d.c. motor can provide a combination of faster controlled acceleration, deceleration and reversal, and precision speed regulation over a wider range. Also, with the control systems and power regulators available for d.c. motor schemes, torque/speed characteristics can be obtained to meet particular control requirements and a very rapid rate of response to control signals can be achieved with motors designed for the purpose.

The inherent torque/speed characteristic of a d.c. motor differs according to whether it has a shunt, series or compound field winding, but the characteristic can be modified considerably by particular arrangements of windings and by regulation of the power supply to the field winding and/or the armature.

D.C.-motor Speed Ranges

The inherent characteristic of the shunt motor is such that the speed drops only about 5% over the normal load range. Regulation of the field current gives a variable speed range of about 5:1, i.e. the top speed is about five times the base speed, which is the rated speed with full field current, the full supply voltage being applied to both the field winding and the armature in parallel. With field-current regulation the torque/speed characteristic of a shunt motor is similar to the constant-power curve shown in *Figure* 4.1. Field control of speed is suitable for constant-power drives, i.e. where the torque required decreases with speed.

To obtain a variable-speed range below base speed, the shunt field is supplied from one source and the armature from another variable-voltage source. Starting from standstill with full field, the armature voltage is increased so that the motor develops constant torque up to base (100%) speed—as shown in *Figure* 4.1. A further increase in speed can be obtained

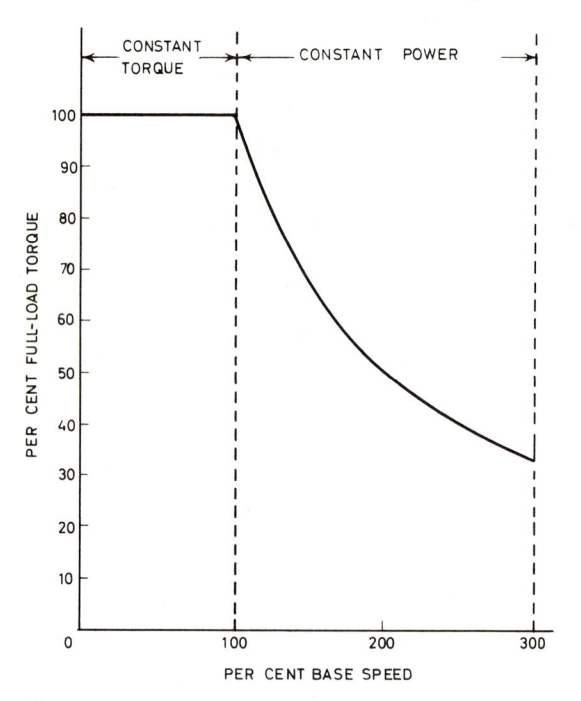

Figure 4.1. Shunt-motor torque/speed characteristic with separate excitation and armature voltage regulation below base speed

by field weakening but with decreasing torque since this is proportional to armature current times field flux.

With armature-voltage regulation the available speed range for constant-torque drives is better than 20 : 1 and for some applications can be as high as 100 : 1.

High-performance Variable-speed Drives

The d.c. shunt wound motor is used extensively for high-performance variable-speed drives, particularly those involving maximum acceleration, deceleration and reversal. With combined armature-voltage/field-excitation speed regulation, maximum acceleration is achieved by starting and runnning up with full field to nearly maximum armature voltage and then obtaining a further increase in speed by field weakening. Subject to current-limit control, the maximum acceleration torque is realised and the run-up time is minimised.

A maximum rate of retardation is achieved by applying regenerative or dynamic braking and maintaining the braking current up to the maximum value permitted by the current-limit control safeguarding the motor against the risk of commutation flashover.

D.C. Shunt Motors for Reversing Duties

To minimise reversal times special types of motors and power regulators are used together with control systems responding very rapidly to a reversing signal input. With such systems large d.c. motors can be reversed from base speed in one direction to base speed in the opposite direction in less than one second. To obtain the deceleration and acceleration required for this rate of reversal the rotating parts of the motor are designed to minimise inertia.

When the motor is reversed by changing the field pole polarities, it is also necessary to maximise the rate of reversal of the field flux. To achieve this the direction of the field excitation voltage is reversed and at the same time it is pushed up to fifteen or more times the normal value so as to force the flow of current through the field winding in a direction that rapidly wipes out the residual pole flux and produces a powerful flux of opposite polarity for reversing armature rotation. For minimising rotational reversal time a flux reversal time of about 0·5 s is essential and to obtain this, in addition to field forcing, the motor yoke must be laminated.

Performance Requirements for Motor Power-supply Unit

When the motor armature is supplied from the generator of a Ward-Leonard set (see page 53) and is reversed by reversing the polarity of the supply to the armature, the rate of reversal of the generator shunt field is a factor determining the rate of reversal of the armature. Consequently, the shunt field residual flux must be eliminated rapidly otherwise it would tend to maintain the generator voltage and therefore the rotation of the motor armature. To neutralise the residual flux the generator has a suicide field which is energised when motor reversal is initiated.

The d.c. exciters supplying the field windings of motors and generators used for reversing drives are also designed for the maximum rate of reponse to control signals which is achieved by design features similar to those adopted for the main machines.

With the later d.c.-motor control schemes, rotary exciters have been superseded by static types so that reversals of excitation current, initiated by electronic controls with their inherent high rate of response, are effected much more rapidly. However, the high response design features of the main machines are still necessary to maximise the rate of reversal when this is a control requirement.

Commutation Requirements

With modern control schemes very high rates of change of armature current also occur and to ensure that the compole flux remains in phase with the armature current, which is necessary to maintain good commutation, the complete magnetic circuit must be laminated.

A rolling mill d.c. motor of 3728 kW (5000 hp) may have to accept simultaneously a rate of change of main-pole flux of four times full-load flux per second, an armature-current change of 60 times full-load current per second, and an acceleration of twice base speed per second. In addition to normal commutation difficulties, the rapid change in main-pole flux may result in a transformer voltage of 1·5 V in each coil undergoing commutation, which cannot be controlled by the compole flux.

The commutation process limits the overload capacity of d.c. motors designed for severe variable-speed duties, e.g. rolling-mill drives. Typically, a motor designed for 100% load at twice base speed has an overload capacity of 225% throughout the constant-torque speed range from zero to base speed which then decreases progressively throughout the constant-power speed range achieved by field weakening.

In practice, it may be impossible to achieve a constant-power output because, with increasing speed throughout the field-weakening range, the torque output falls below the value necessary for constant power because

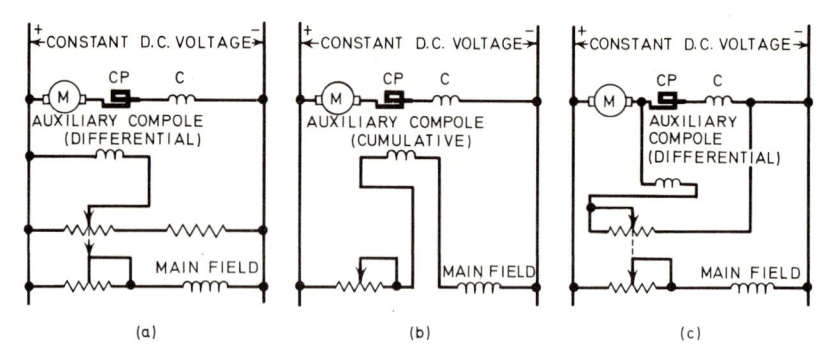

Figure 4.2. Auxiliary compole windings
M : motor armature   CP : compole winding   C : compensating winding

of the limit imposed by the commutation process. To maximise the overload capacity of variable-speed d.c. motors, compensating windings may be used in addition to the normal compole windings. Both these windings are excited as a direct function of the armature current.

In non-compensated motors the strength of the commutating field is usually insufficient on heavy peak loads because of saturation of the compole magnetic circuit. By providing compensating windings to overcome the effects of armature reaction, the overload at which saturation becomes limiting is greatly reduced.

With variable-speed motors the correct compole field strength for optimum commutation performance varies with speed owing to the increase in eddy currents in the armature copper at the higher frequencies and the corresponding reduction in the cross-slot flux, particularly when the armature has relatively deep conductors. The effective reactance of the armature coils undergoing commutation changes with frequency so as to require a smaller commutating flux at the higher speeds. This reduction may result in 3–5% less compole ampere-turns being required at top speed on a motor with a 4 : 1 speed range. Over-strong compoles have a secondary effect on speed regulation and inherent stability at top speed, and motors with a wide speed range may be fitted with auxiliary compole windings to overcome this disturbance.

From tests carried out by AEI on a 2498 kW (3350 hp), 500 V, 300–750 rev/min d.c. motor to determine the limits of the bandwidth of compole field strength outside which visible sparking occurs, it was seen that at 750 rev/min the required mean compole field was 3% less than at 300 rev/min. Thus a compole field strength directly proportional to load only would not give the correct setting over the full speed range.

In many cases sufficient compensation is effected by a constant value of differential ampere-turns at all loads, but in others it is necessary to introduce a differential excitation which is proportional to load and is also a function of speed. An alternative is for the compoles to be adjusted for optimum commutation at top speed and to use auxiliary coils arranged to boost the main compole coils at the lower speeds.

Auxiliary Compole Windings

Figure 4.2 shows three methods used by AEI for applying adjustment for speed by means of an auxiliary compole winding. In *Figure* 4.2(a), mechanically coupled rheostats are operated to weaken the auxiliary compole differential field as the main field is strengthened, and vice-versa. In *Figure* 4.2(b) the auxiliary compole winding is connected cumulatively and in series with the main field, its strength being arranged so that at weak main field the correct reduction in commutating field is obtained. With these two methods the compole field is not a function of load, but both are satisfactory on motors for which commutation is not critically difficult.

With the third method, *Figure* 4.2(c), the compole field is a function of load as well as of main field strength. Coupled rheostats are used as in the first method but, instead of the potentiometer rheostats shown, if desired contactors can be included to close at the higher speeds. For severe duties the auxiliary coils can be supplied by an amplidyne exciter (see page 54) or other d.c. amplifier which is made sensitive to speed from the output of a tachogenerator and to load by tappings across the compole and compensating windings.

D.C. Series Motor

With the series-wound motor the starting current flows through both the armature and the field winding, so that the starting torque is a maximum when the current is limited to the maximum value for which the motor is designed. With rising speed the back e.m.f. increases and the current and torque decrease. The operating speed rises or falls with decreasing or increasing load, respectively. The series motor is used for crane and hoist drives and for certain industrial haulage drives.

Speed regulation can be obtained by a series resistor or by a continuously variable or tapped resistor connected across the series field winding as a diverter.

D.C. Compound Motor

The compound motor usually has a cumulative series winding which with increasing load increases the total field flux so that the motor slows down. The main application of cumulatively compounded motors is for drives incorporating a flywheel with which it is necessary to obtain a sufficient drop in speed to enable the flywheel to release stored energy during heavy load peaks in order to avoid excessive overloading of the motor.

For certain special applications a compound motor has a differential series field which with increasing load reduces the total field flux so that the motor speeds up.

Motors with both cumulative and differential series windings are used with the four-roll bridles of a strip electrolytic tinning line, equipped with AEI main electrical drives, to ensure that the load is shared correctly between the motors driving the individual rolls. Load sharing becomes critical when the bridle reaches its maximum tension capacity and the slip points of the rolls are reached. Any tendency for one drive to pick up more load may result in slipping at the roll so that extra load is imposed on the other rolls.

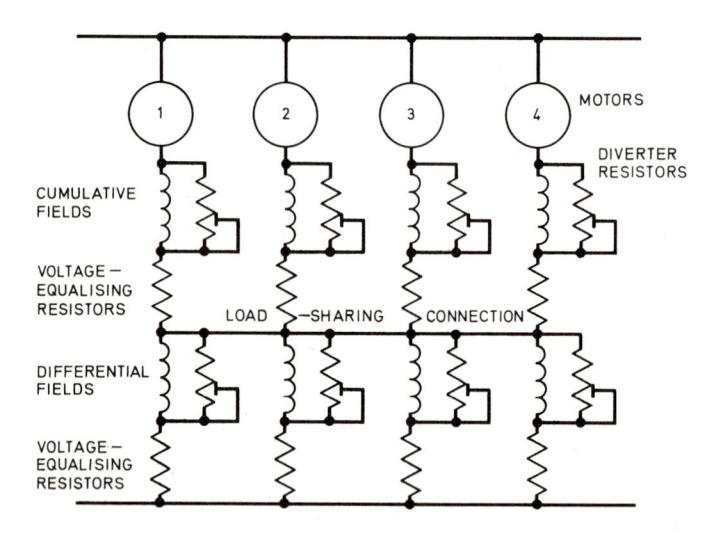

Figure 4.3. Arrangement of compound-wound motors for bridle drive (Shunt fields are not shown)

Slipping then occurs on these rolls and there is a loss of tension. With four-roll bridles, very close control of load sharing is necessary to avoid slipping.

To meet the control requirement for the bridles of the electrolytic tinning line, AEI used the scheme shown in *Figure* 4.3 to restore load balance quickly by direct action both on the overloaded motor and on the other

D.C. Motors and Power Regulators

three motors which must take more load to maintain the overall bridle tension.

Each motor has equally effective cumulative and differential fields. Operating in conjunction with the shunt field, the cumulative field would give an inherent speed regulation producing a 50% drop in speed from no-load to full-load, but this is cancelled out by the action of the differential field. The exact torque/speed characteristic is established by adjusting the

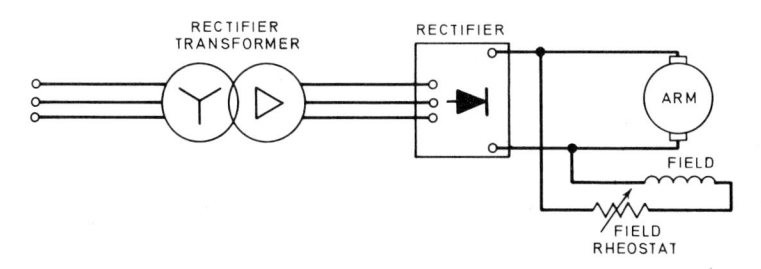

Figure 4.4. Shunt-field regulator speed-control scheme with rectifier power supply

diverter resistors in parallel with the series fields. If one motor becomes overloaded the current strengthens its cumulative field and also the differential fields of all the motors. The overloaded motor slows down to shed load and the other motors speed up to take load. The greater the effect of the series fields, or the greater the change in speed due to a change in load, the more accurate will be the load sharing.

D.C. Shunt-motor Power Regulators

A shunt motor may have a series stability winding to avoid instability due to the weakening of the field by armature reaction.

The simplest shunt-motor power regulator is a rheostat in the field circuit (*Figure* 4.4), and for automatic control it can be driven by a pilot motor. Alternatively, the shunt field can be supplied separately from a static or rotating exciter which can be controlled to give a variable-voltage output.

For the independent regulation of the power input to the armature either a motor-generator or some form of rectifier unit is used. Both these d.c. sources may be used to provide a variable-voltage armature supply to several motors when they have to be controlled as a group (see *Figure* 4.3) otherwise each motor is supplied individually from a separate generator or rectifier, which can be controlled automatically to function as a power regulator.

Ward-Leonard Power Regulators

The Ward-Leonard set is a combination of a motor-generator and a motor forming a self-contained drive using any suitable type of a.c. motor coupled to the generator which supplies the motor armature. The generator and motor shunt fields are excited from a separate source. Power regulation is obtained by varying the generator field excitation, or the generator and motor fields in sequence if the widest speed range is required. The basic scheme is simple but it has been developed to include control and regulation features that ensure more rapid response, closer speed holding over an extended range, and current limiting during acceleration and deceleration, which enables the rates to be maximised.

The excitation system of the Ward-Leonard generator is usually specially designed to meet the particular control requirements of the drive. With many drives it is essential to control the effects of the residual field flux, which persists after the excitation current has been reduced to zero. This flux is strong enough to generate sufficient current to keep the motor armature rotating at crawling speed while the motor field is excited.

If the requirement is simply that armature rotation must be stopped quickly, the residual flux can be neutralised by a flux produced by a suicide field winding connected to the generator terminals when its excitation is reduced to zero. To enable the motor speed to be regulated precisely from zero to base speed, a bias winding in the generator field system is used to neutralise the residual flux (see page 84).

To minimise motor reversal time a suicide field may be used to kill the residual flux and initiate flux reversal while the main generator field is being reversed; or there may be two field windings—one for each direction of rotation.

With some Ward-Leonard schemes rapid response to control signals is achieved by energising the generator field from a rotating exciter with a field system supplied by a rotating amplifier exciter giving an output that changes rapidly with the variation of a signal input to its control field winding. To minimise the overall response time of the scheme the motor field may be energised from a similar excitation system, and to achieve rapid motor reversal the exciter can be designed to force the main-field excitation voltage up to several times its steady-state value.

Amplifier exciters are also designed with control windings for specific control functions, e.g. to limit the armature current of the Ward-Leonard generator during acceleration and regenerative braking of the drive motor.

Rotating-amplifier Excitation

Figure 4.5 shows a simplified speed-regulating scheme with an amplidyne rotating amplifier supplying the generator field winding. The amplidyne is driven at constant speed by the motor driving the generator.

The output of the amplidyne depends on the input to its control winding which is determined by the error voltage, i.e. the difference between the reference voltage *bc* and the tachogenerator voltage *bd*. Under steady-state conditions the error voltage is just sufficient to give the amplidyne output necessary to obtain the generator excitation required to maintain the motor speed set by adjustment of the potentiometer R. A change in speed alters the error voltage, the amplidyne output and the generator excitation.

With the simple scheme shown the amplidyne has an additional field winding which is energised by a stabilising voltage feedback whenever there

D.C. Motors and Power Regulators

is a sudden change in speed. This feedback serves to prevent the motor hunting due to over-correction of the error voltage causing the speed to oscillate above and below the required value. In effect, the stabilising feedback minimises the time taken for the motor speed to recover to the set value.

At one time rotating amplifiers were used effectively in many Ward-Leonard schemes, but it is now a common practice to use thyristor converters to supply and control excitation systems.

With some Ward-Leonard schemes a small booster generator connected in series with the main generator and motor armatures (see page 74) is used

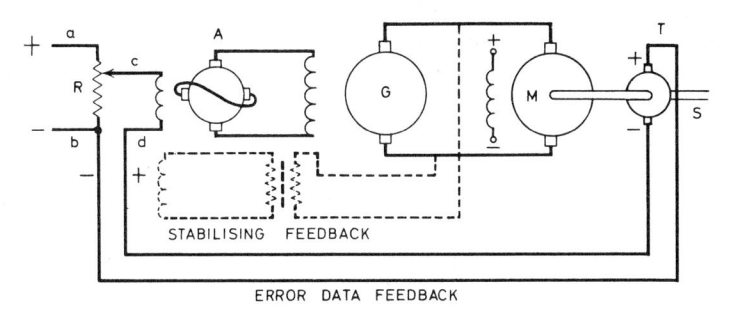

Figure 4.5. Ward-Leonard set with amplidyne generator field exciter

to meet particular speed-regulation requirements for certain drives. The booster generator may boost or buck the main generator voltage.

With the introduction of various types of rectifier units fewer generator-fed d.c.-motor drives are being used, but a considerable number are, of course, still in operation. Similarly, the various valve-type rectifiers which were used as alternatives to motor-generator sets are now being largely superseded by semiconductor units.

Static Power Regulators

In the past, thermionic, thyratron, excitron, ignitron and mercury-arc rectifiers have been used to provide variable-voltage supplies for d.c. motors with ratings within the capacity of each particular type of rectifier, but the present tendency is towards the exclusive use of the thyristor converter.

Mercury-arc Converter Power Regulators

Developed as a replacement for the Ward-Leonard set for high-power heavy-duty drives, the mercury-arc converter continues to operate in many installations but for the same duties it has been superseded by the thyristor converter.

As with the thyristor converter (see page 58), the output of the mercury-arc converter is regulated by varying the conduction period during the positive half-cycles of the voltage applied to the anodes.

The basic components and circuits of a 3-phase converter are shown in *Figure* 4.6. Between each anode and the cathode is a grid to which voltages are applied by the grid control gear to establish the instant at which the anode begins to conduct. The anode can be held off by negative bias voltage applied to the grid continuously, and conduction is initiated by applying a steep-fronted positive voltage that swamps the negative voltage.

When the positive voltage is applied to the grid at the beginning of the positive half-cycle of anode voltage, firing occurs naturally as the anode voltage rises. In this case, each anode of a 3-phase converter conducts in turn and the voltage applied at the cathode has the waveform shown by the heavy line in *Figure* 4.7(a). The average voltage at the cathode is shown by the dotted line and it is at a positive potential relative to the d.c. load.

The instant at which an anode fires can be controlled by delaying the application of the positive grid voltage, in other words, by retarding the

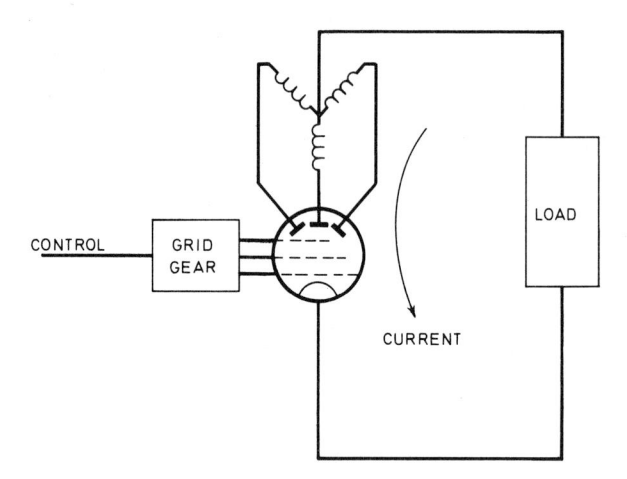

Figure 4.6. Schematic of 3-phase mercury-arc converter

grid firing angle. Stated in electrical degrees, this angle establishes the instant in the anode positive-voltage half-cycle when the anode begins to conduct. Referring to *Figure* 4.7(b) it will be noted that although firing occurs at 90° relative to the commencement of the positive half-cycle, the firing angle is given as 60° because it is measured from the instant (30°) when each anode would have fired if it had not been held-off.

Since the effect of retarding (or 'phasing back') the grid firing angle is to reduce the conduction period, the average d.c. potential at the cathode and the output of the converter are also reduced. With the 60° firing angle the d.c. voltage applied to the load is reduced to 50% of the full voltage available from the converter.

In *Figure* 4.7(c) the firing angle is 90° so that the average cathode potential is zero and is therefore at the same potential as the neutral point of the transformer supplying the converter. The negative terminal of a d.c. motor is connected to the transformer neutral point.

Beginning with a firing angle retarded to 90°, the starting current of a d.c. motor can be regulated by gradually advancing (or 'phasing forward') the firing angle, and acceleration and running speed can be regulated by

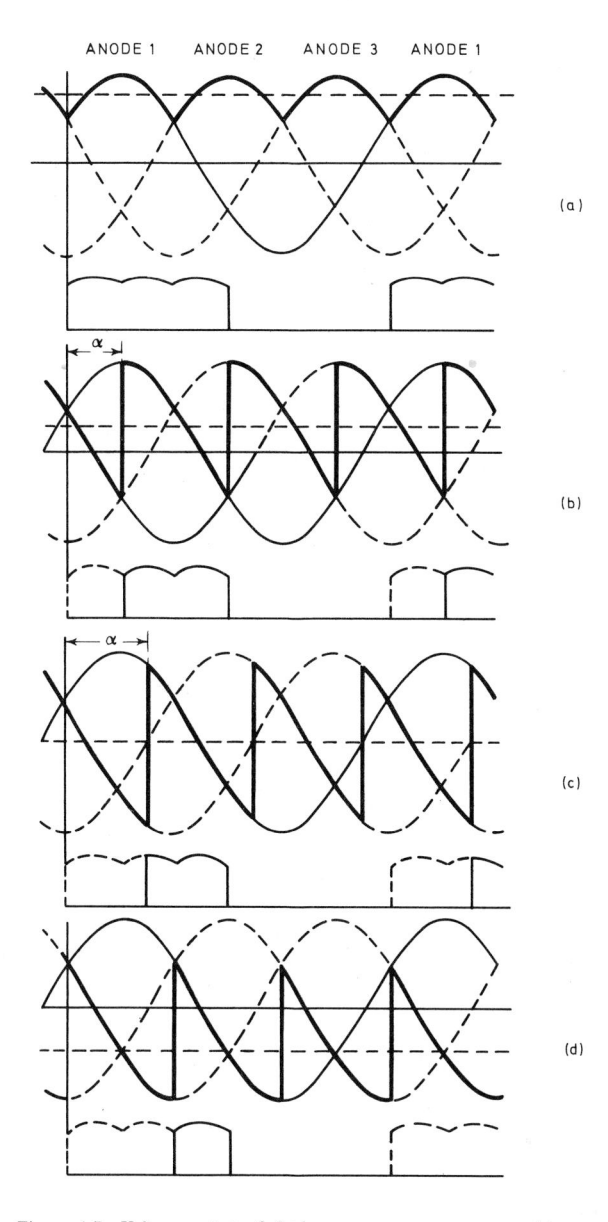

Figure 4.7. Voltage output of 3-phase mercury-arc converter with variation in firing angle
(a) Rectifier d.c. voltage maximum; zero grid delay ($\alpha = 0°$);
(b) Rectifier d.c. voltage reduced to 50%; 60° grid delay ($\alpha = 60°$);
(c) Rectifier d.c. voltage reduced to zero; 90° grid delay ($\alpha = 90°$);
(d) Inverter voltage now 50% of maximum; 120° grid delay ($\alpha = 120°$)

Below the anode 1 applied voltage waveform (full line) is the positive grid voltage waveform shown as a full line when it is applied to enable anode 1 to conduct control signals initiating the instant of the application of the positive grid voltage by the grid control gear.

Converter-fed Reversing Drives

The mercury-arc converter provides an infinitely variable voltage to the armature of a d.c. motor for speed regulation, but as it conducts only from anode to cathode, a converter-fed reversing drive must include switchgear to reverse either the motor armature or field connections, or alternatively, the motor armature or field can be supplied from two converters—one for each direction of rotation.

When the time taken does not have to be minimised, reversing can be carried out by first retarding the firing angle to stop the current flow; disconnecting the armature from the converter and connecting it to a resistor for dynamic braking; reversing the armature or field connections; disconnecting the armature from the braking resistor and re-connecting it to the converter; and advancing the firing angle to give a current output for accelerating the motor in the opposite direction.

With this sequence of operations reversal time is too long to allow for effective control of certain drives requiring a minimum dead time, i.e. the time when with no current flow the motor is not developing torque so it is not under control. Consequently, during the dead time no control is exercised over the driven unit or the operation it is performing. For example, loss of control of a strip processing line may result in a departure from the required tension values breaking the strip.

Various mercury-arc converter (and thyristor-converter) schemes developed to minimise dead time involve regenerative braking of the motor while it is still connected to the converter. To enable the motor to regenerate into the supply, the converter voltage must be phased back until, first, the current flow is stopped and then the average cathode potential becomes less than zero, i.e. below the potential of the transformer neutral point. *Figure* 4.7(d) shows the anode-voltage waveform when the grid firing angle has been phased back to 120° and the converter can function as an inverter to allow the motor to regenerate into the supply when either its armature connections or the field flux has been reversed. When this has been done, the motor positive terminal is connected to the neutral point of the transformer (*Figure* 4.6) and the generated current flows from the anode to the cathode in the normal way.

By variation of the grid firing angle, precise and rapid regulation of the current flow is achieved when the converter is either rectifying or inverting so that, except during the dead time, the motor is under close control and can be accelerated and retarded by regenerative braking at the maximum rates permitted by a current-limit control. Automatic control of both the motor and braking current is effected by signals initiating adjustment of the grid firing angle in accordance with the varying control requirement for the operation of the driven unit.

Thyristor Converter-fed Drives

The thyristor converter is now being adopted for d.c.-motor drives previously supplied by the mercury-arc converter. Both types of converter have similar control characteristics, the rectifying and inverting modes of the thyristor also being controlled by variation of the firing angle (also referred to as the 'trigger angle'). As with the mercury-arc converter, motor reversal is obtained by reversing the armature connections or the field flux, or by using two converters. The basic reversing schemes for both types of converter are therefore similar in principle so that the thyristor converter schemes shown in *Figures* 4.8–4.10 also resemble those used for the mercury-arc converter reversing drives still in service.

Armature-reversal Scheme

Figure 4.8(a) shows an armature-reversal scheme with high-speed contactors to change-over the connections. The contactors are not required to break any appreciable current or voltage since any arcing at the breaking contacts might persist as the making contacts close thereby short-circuiting the converter and the motor.

The firing angle is retarded to stop the current flow before the contactors can operate to change-over the connections.

As shown in *Figure* 4.8(a) for forward motoring, the converter cathode is positive and current flows to the motor positive. For forward regenerating the cathode is negative and with the armature connections reversed the motor positive is connected to the anode. For reverse motoring the converter voltage is then phased forward to give a current input to reverse and accelerate the motor with the converter operating as a rectifier. With this scheme, dead time is typically 0·4 s.

Field-reversal Scheme

With the field-reversal scheme shown in *Figure* 4.8(b) the motor field winding is supplied from either one of two auxiliary thyristor converters connected in inverse-parallel and with firing control circuits giving rapid changeover from one field converter to the other as required by the control sequence for the main converter. To obtain rapid reversal of the flux, the field converters are rated to apply a forcing voltage much higher than the normal value.

The control procedure is similar to that for the armature-reversal scheme. The main converter voltage is phased back to reduce the armature current to zero and to make the cathode potential negative, the conducting field converter voltage is phased back to zero and the other converter is fired to give the forcing voltage required to reverse the flux rapidly and to enable the motor to regenerate (*Figure* 4.8(b)). For reverse motoring the conducting field converter voltage is adjusted to give normal excitation current. The dead time is typically 0·5 s but varies according to the amount of field forcing that can be applied economically.

Double-converter Reversal Schemes

Dead time during reversal can be further reduced by using two converters, one for each direction of rotation, cross-connected as shown in *Figure* 4.9. With this back-to-back arrangement, the converters are controlled so that either can rectify or invert. Since with this arrangement no mechanical

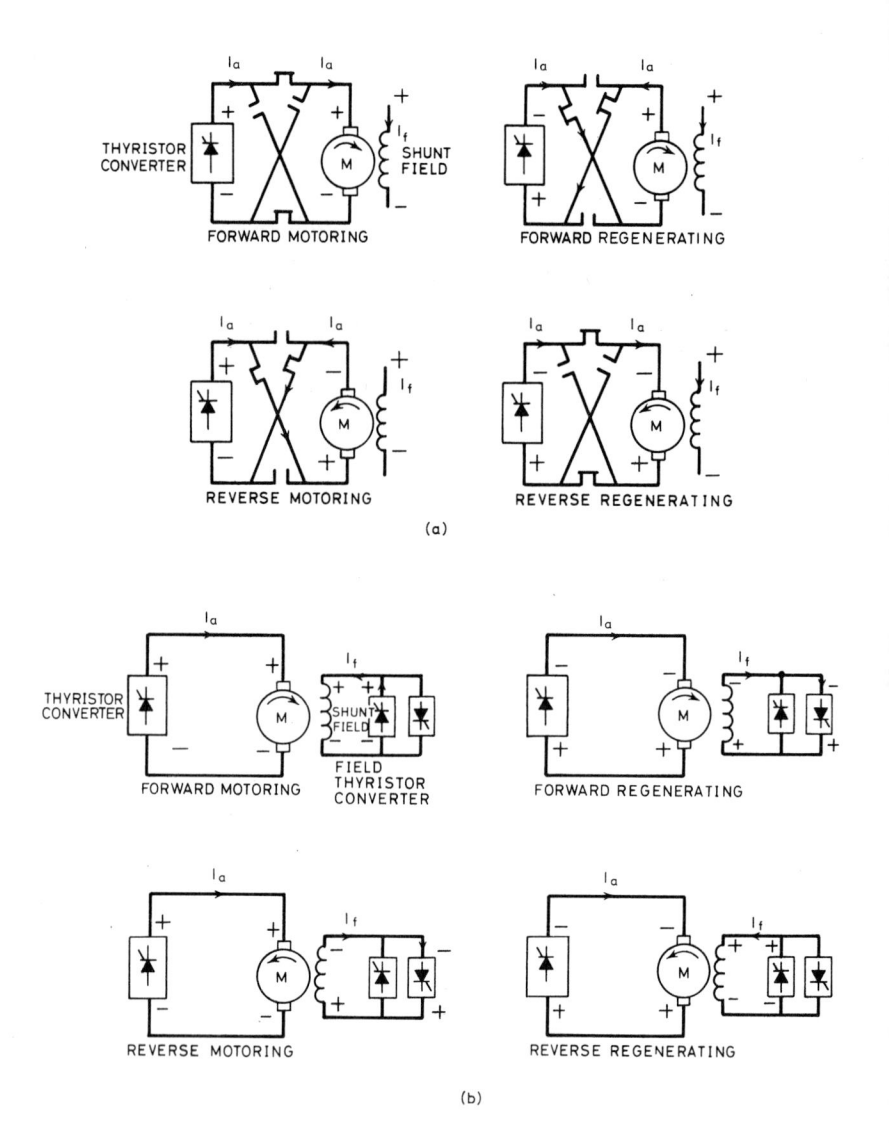

Figure 4.8. Thyristor converter basic figure-0 reversing drive using (a) armature connection reversing contactors and (b) field reversal by a double-converter system

D.C. Motors and Power Regulators

switching or flux reversal is necessary, if the control scheme ensures that the converter voltages are always equal and opposite, the change from drive to regeneration torque can be completely smooth with no hesitation at zero current. The rate at which current can be reversed is limited only by motor commutation considerations.

However, although the converters can be controlled so that the mean inversion voltage of one always equals the mean rectifying voltage of the other, the instantaneous voltage values are not always equal owing to slight differences in the voltage waveforms. This results in a circulating current around the loop formed by the converters. There are certain disadvantages

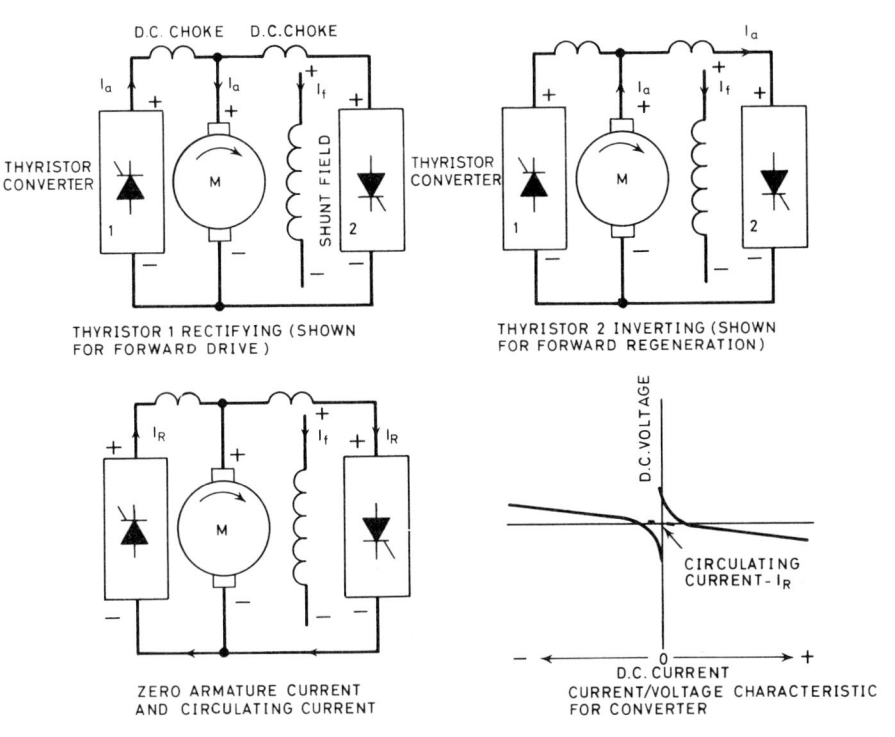

Figure 4.9. Thyristor converter figure-8 back-to-back scheme for reversing drives and armature voltage regulation

with schemes involving circulating current but for drives where scarcely any dead time at all can be tolerated, the current can be limited to an acceptable value by using a transformer with two isolated secondary windings, one for each converter, and special d.c. chokes.

With the scheme shown in *Figure* 4.10 a transformer with a single secondary winding supplies both converters, and there are no special d.c. chokes, circulating current being avoided by allowing only one converter to conduct at any instant. A logic circuit (see page 68) determines which converter conducts at any instant and prevents changeover from one converter to the other until current flow has ceased. The changeover control is similar to that used to initiate operation of the reverser in the armature-reversal scheme but the changeover is much faster, the dead time being only a few milliseconds.

Continuing development of thyristors and control schemes provides for precisely regulated power supplies for d.c. motors in ratings up to several thousand kilowatts. Thyristor power-supply and control units are now available in ranges for a variety of industrial drives, or basic modules can be used to engineer supply and control units incorporating different facilities to meet the control requirements of particular drives. An inherent feature of

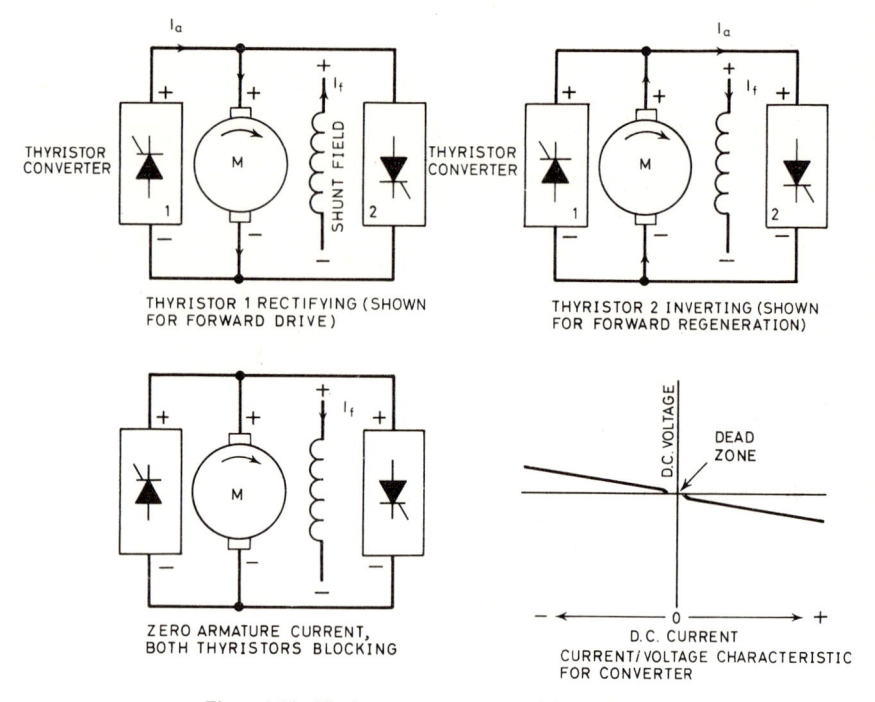

Figure 4.10. Thyristor converter suppressed figure-8 scheme

thyristor converter schemes is the wide range of power regulation available—typically from less than 0·25% to greater than 99·9% of maximum power.

Types of Thyristor Converters

For certain drives the motor armature supply is obtained from a unit incorporating a combination of thyristors and diodes which meets the control requirements and is cheaper than an all-thyristor unit.

Figure 4.11 shows the connections of a thyristor/diode mixed bridge converter which is used for small d.c. motors requiring only a small speed-control range.

The all-thyristor fully controlled bridge converter shown in *Figure* 4.12 operates as a rectifier or inverter, gives speed regulation down to zero and is used for reversing drives with field or armature reversal switching.

D.C. Motors and Power Regulators

The mixed bridge converter shown in *Figure* 4.13, comprising a thyristor bridge in series with a diode bridge, is similar to that shown in *Figure* 4.11 but is used for larger d.c. motors with a limited speed range.

The anti-parallel thyristor bridge converter shown in *Figure* 4.14 is for reversing motor drives with full speed control in both directions without switching motor connections.

Diode-rectifier Power Regulators

Rectifier units incorporating semiconductor diodes are used to supply d.c. motor drives when the control requirements are such that adequate speed regulation can be obtained by adjusting the voltage applied to the field or/ and the armature by regulating the a.c. voltage input to the rectifier.

A rectifier transformer with on-load tap-changing provides for speed

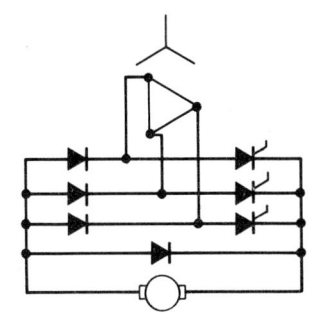

Figure 4.11. Thyristor/diode bridge

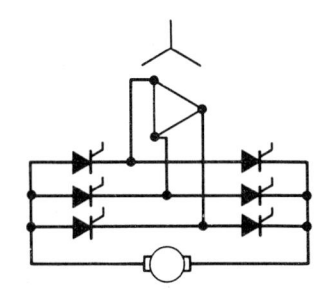

Figure 4.12. Fully controlled thyristor bridge

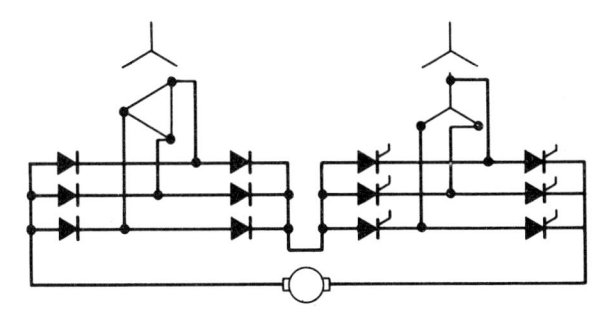

Figure 4.13. Mixed bridge with thyristors in series with diode bridge

variation in definite steps, but some type of infinitely variable a.c. voltage regulator is generally used for variable-speed drives.

Figure 4.15 shows a basic rectifier-fed d.c.-motor scheme with a.c. voltage regulation of the input to the field rectifier and a separate constant-voltage rectifier for the armature. Automatic speed control is by the tachogenerator output to a relay controlling the operation of the field supply regulator. A similar scheme for a variable-voltage d.c. armature supply is shown in *Figure* 4.16.

For certain low-power d.c. motor drives the rectifier is supplied directly from the a.c. medium-voltage mains.

The input to a rectifier can also be regulated by a static transductor (saturable reactor), which is useful for certain automatic control schemes since its a.c. output is regulated by a variable d.c. voltage applied to a

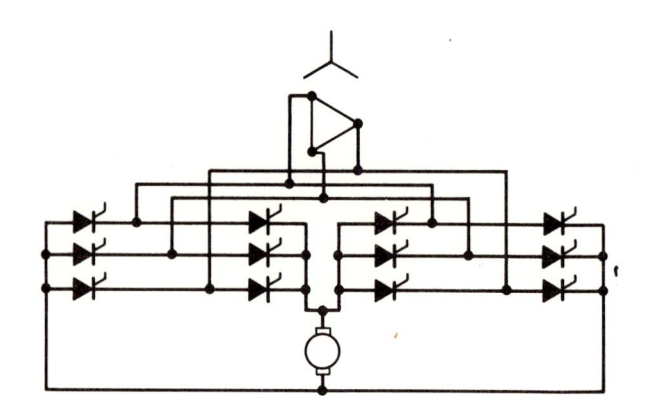

Figure 4.14. Anti-parallel connected thyristor bridges for reversing drives

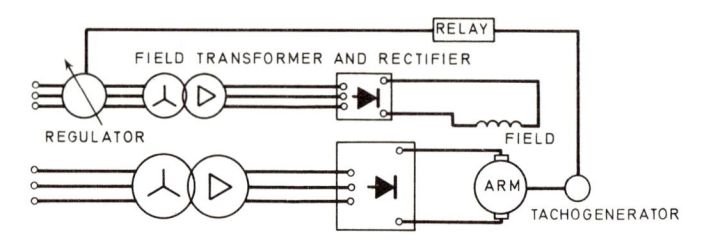

Figure 4.15. Diode-rectifier-fed drive with speed control by a.c. voltage regulation of field power supply

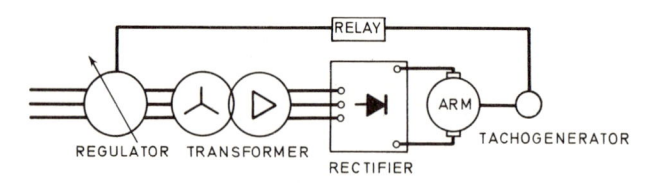

Figure 4.16. Diode-rectifier-fed drive with speed control by a.c. voltage regulation of armature power supply

control winding. As shown in *Figure* 4.17 a simple transductor has two cores with an a.c. and d.c. winding on each. The d.c. windings are connected in series opposition and the a.c. windings may be connected either in series or in parallel. Adjusting the d.c. voltage input to the control winding varies the direct current, the saturation of the core material, the impedance of the a.c. winding and therefore the current in the load circuit. The impedance decreases as the control-winding current increases.

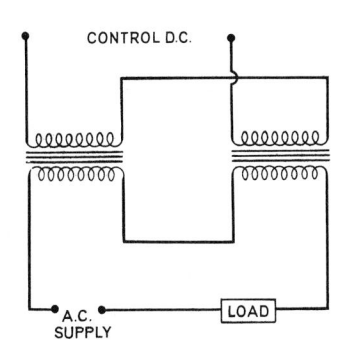

Figure 4.17. Basic arrangement of transductor windings

Transductor power regulation is used in conjunction with transformer tap-change gear to provide adjustable reduction of voltage for inter-tap control. The application of the transductor is limited by the fact that its size is directly proportional to the amount of voltage reduction required, and by the effect on power factor and waveform. It is suitable only if the control requirements do not include the very high speed of response, or inverted operation, available with controlled rectifiers.

Chapter 5

Control Devices and Circuits

In operation an automatic drive is controlled by signals from the various devices which ensure that the motor power input is regulated so as to achieve the required operation of the driven unit and safeguard this and the electrical equipment against adverse conditions such as overload and overspeed.

At some time, however, an automatic drive has to be started, stopped and perhaps regulated by manual control action so a control scheme includes devices for these purposes. The basic devices for manual control are included in the typical circuits shown in *Figure* 5.1(a)–(d) but for automatic control the scheme will necessarily include additional devices.

Automatic Switching Devices

For automatic on/off control, the circuit shown in *Figure* 5.1(e) includes a remote pilot switch (13), e.g. a float switch, pressure switch or thermostat, and also a rotary switch (12). In the position shown the rotary switch is in series with the pilot switch (13), but it can be moved to a position for use as a hand-operated on/off switch.

To ensure safe and correct operation of a drive, the control scheme may include various switching devices in circuit with the operating coils of relays and auxiliary contactors controlling the main power supply switching and regulating equipment. For example, there may be two limit switches in series which have to be closed to energise a relay with contacts which close to complete the supply circuit through a 'start' switch to the operating coil of the main contactor switching the supply to the motor. If one or both limit switches are open, the main contactor cannot be closed. Other schemes include various control-circuit switching devices with relays and contactors for sequencing and interlocking main switching operations. Interlocking is essential with schemes where one control action must be completed before another can be initiated.

Static Switching Systems

Complex control schemes may include a considerable number of relays and contactors and as these are electromagnetic contact devices, preventive

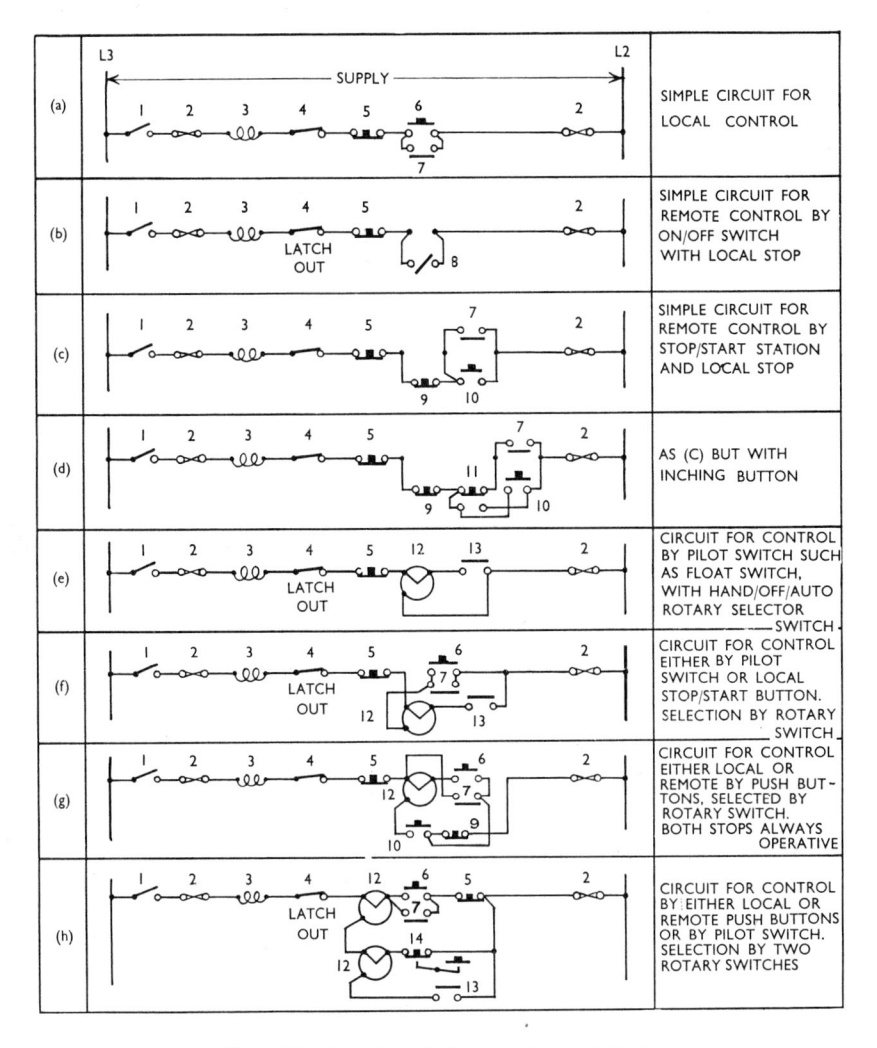

	Circuit	Description
(a)	L3 — SUPPLY — L2; 1 2 3 4 5 6 2, 7	SIMPLE CIRCUIT FOR LOCAL CONTROL
(b)	1 2 3 4 5 2; LATCH OUT 8	SIMPLE CIRCUIT FOR REMOTE CONTROL BY ON/OFF SWITCH WITH LOCAL STOP
(c)	1 2 3 4 5 7 2; 9 10	SIMPLE CIRCUIT FOR REMOTE CONTROL BY STOP/START STATION AND LOCAL STOP
(d)	1 2 3 4 5 7 2; 11 9 10	AS (C) BUT WITH INCHING BUTTON
(e)	1 2 3 4 5 12 13 2; LATCH OUT	CIRCUIT FOR CONTROL BY PILOT SWITCH SUCH AS FLOAT SWITCH, WITH HAND/OFF/AUTO ROTARY SELECTOR SWITCH
(f)	1 2 3 4 5 6 2; LATCH OUT 12 7 8 13	CIRCUIT FOR CONTROL EITHER BY PILOT SWITCH OR LOCAL STOP/START BUTTON. SELECTION BY ROTARY SWITCH
(g)	1 2 3 4 5 6 2; 12 7 9 10	CIRCUIT FOR CONTROL EITHER LOCAL OR REMOTE BY PUSH BUTTONS, SELECTED BY ROTARY SWITCH. BOTH STOPS ALWAYS OPERATIVE
(h)	1 2 3 4 12 6 5 2; LATCH OUT 7 12 14 13	CIRCUIT FOR CONTROL BY EITHER LOCAL OR REMOTE PUSH BUTTONS OR BY PILOT SWITCH. SELECTION BY TWO ROTARY SWITCHES

Figure 5.1. Manual and simple automatic control circuits
1. Aux. contact on isolator
2. Control circuit fuses
3. Main contactor coil
4. Overload trip switch
5. Local 'stop' button
6. Local 'start' button
7. Retaining contact
8. On/off rotary switch
9. Remote 'stop' button
10. Remote 'start' button
11. Inching 'start' button
12. Rotary selector switch
13. Remote pilot switch e.g. float, pressure or thermostat
14. Remote lock-out 'start/stop'

maintenance is costly if they are operating more or less continuously with drives for repetitive duties. The operating time of electromagnetic devices is also relatively long so that with a scheme using a series of devices the response to a control signal is slow.

For many control schemes with a minimum response time requirement, a static switching system is an economic alternative to the use of electromagnetic devices. Typical systems use transistor and diode gates which are now capable of consistent reliability with minimal maintenance and operate about 1000 times faster than electromagnetic devices.

In effect the semiconductor gates operate as contactless switches performing so-called logic functions. *Figure* 5.2 shows schematically the gates for the logic functions AND, OR and NOT together with the equivalent

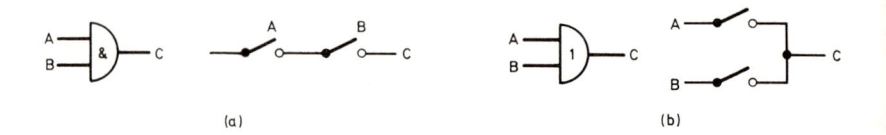

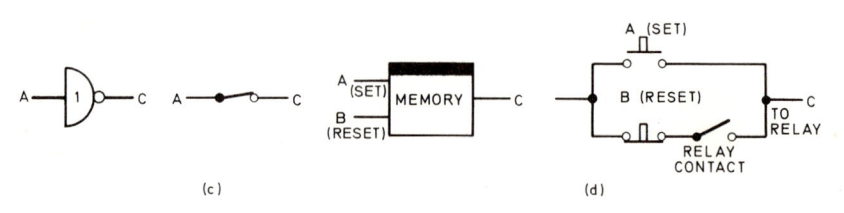

Figure 5.2. Static switching system units

relay contacts for the same function. The AND gate (*Figure* 5.2(a)) gives an output at C when there is an input at both A and B. It is equivalent to two relays each of which has to be energised by a control signal to close the contacts A and B in series to give an output at C. The OR gate gives an output at C when there is an input at either A or B, or both, and is equivalent to two relays in parallel (*Figure* 5.2(b)). With the NOT gate there is an output at C when there is no input at A, i.e. an input at A stops the output at C so that the gate is equivalent to a relay with contacts closing the circuit when de-energised, and opening the circuit when it is energised by a control signal (*Figure* 5.2(c)).

In addition to the basic logic gates, a static switching system also includes units for such functions as 'memory', 'trigger' and 'time delay'. The memory unit shown in *Figure* 5.2(d) gives an output at C when there is an input at A (Set) and the output is maintained even when the input signal is removed. The output is stopped when an input is applied to B (Reset). With the equivalent relay circuit, operation of the pushbutton A energises the relay to close the contact which maintains the output when the pushbutton is released. The output is stopped when the pushbutton B is operated to de-energise the relay and open the contact.

A trigger unit gives an output only when an input signal is present and it can be turned off by applying another input. A time delay operates similarly, except that the output appears only some definite time after the input signal has been applied.

Various other units are included in the different static switching systems now available for complex control schemes using voltage signals from a wide range of devices.

It will be appreciated that static switching is not necessarily advantageous for control schemes when rapid response to control signals is not essential, and the scheme includes simple contact devices, e.g. limit switches, which can be used directly in motor-control circuits. An advantage of static switching units is that they accept low-voltage inputs available directly from sensing units such as proximity and photoelectric detectors so that control schemes can be designed without any contact devices, consequently there is a minimum time lag between the demand for and the initiation of control action. The response of a control scheme is especially fast when the output of the static switching system is used for the operation of a double-converter power unit.

For some drives the output of the static switching system is used for the operation of electromagnetic contactors so the system may include power amplifiers to raise the output to the appropriate level.

Continuous Speed-control Schemes

For simple on/off control, the starting and stopping of the motor is initiated by devices that establish lower and upper limits of the magnitude of the controlled quantity that must not be exceeded. If the controlled quantity has to be maintained close to a desired value and it tends to fall below the lower limit at short intervals, this involves frequent switching of the motor.

When the controlled quantity has to be maintained close to a desired value with varying output, the motor can be run continuously and its speed regulated in accordance with the demand. For example, the speed of a motor-driven pump can be regulated, to maintain the level in a storage tank supplying a varying demand, according to the rate at which the level tends to fall, or the flow through a pipeline tends to change.

With control schemes intended to initiate corrective action more or less continuously as the measured value of the controlled quantity deviates from the desired value, this is represented conveniently by a voltage which is an analogue of whatever parameter is being measured. The present discussion is concerned with the control of motor speed.

Speed Reference Setting Devices

Analogue speed reference voltages are set by a potentiometer connected to a stabilised power-supply unit. A typical unit provides a 200 V d.c. output maintained to within $\pm 0\cdot 02$ V for long periods over a temperature range of $25°C$ and mains voltage variations of $\pm 10\%$.

When the absolute accuracy of speed setting is not important, and small incremental speed adjustments are required, coarse-and-vernier speed-setting potentiometers are used.

When precise speed setting is required to an absolute accuracy of the order of 0·1% of maximum speed, with incremental settings of 1 rev/min, a bank of three or four low-contact-resistance decade switches can be utilised in conjunction with an equal number of banks of suitably graded high-precision wire-wound resistors. These resistor banks are connected across the stabilised voltage supply unit and arranged such that the speed reference voltage can be set in terms of revolutions per minute. For speed ranges of 0–1000 rev/min, three decade switches would provide for the selection of a desired speed value in increments of hundreds, tens and units.

Motor-driven potentiometers are used to provide for remote-controlled setting, and for the automatic regulation of the motor acceleration and retardation, desired values of these being selected by a control adjusting the speed of the potentiometer motor (see page 111). Acceleration and retardation are regulated in accordance with the progressive variation of the reference

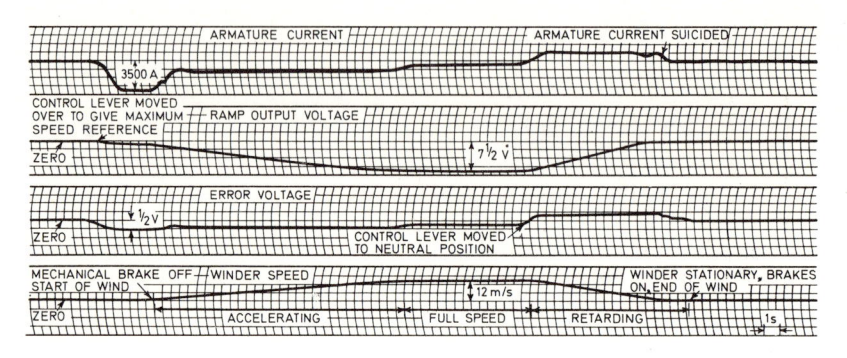

Figure 5.3. Curves of winder speed, error voltage, ramp generator output voltage and armature current

voltage with time, i.e. the voltage/time characteristic of the motor-driven potentiometer.

Ramp-function Generators

Voltages varying with time can also be produced by an electronic ramp-function generator (usually referred to simply as a ramp generator), which is used with static control schemes for precise regulation of speed during acceleration and retardation to enable maximum rates to be achieved safely. By maximising these rates, the times to run up to operating speed and brake to standstill are minimised, which is an advantage when these times form a large proportion of the overall time of a duty cycle.

The control exercised by a ramp-function generator is demonstrated by the performance curves shown in *Figure* 5.3 obtained during commissioning trials of an AEI Ward-Leonard drive for a colliery cage winder. The duty cycle of the winder is shown in *Figure* 2.1.

For the control of the winder, the driver has a desired speed reference potentiometer, which is set to provide a control signal to initiate operation of the winder when the driver's control lever is in the forward or reverse position. The ramp generator consists basically of a transistorised integrator unit which converts step input voltages from the driver's reference potentiometer to a voltage output which rises linearly with time, corresponding to the normal acceleration time of the winding cycle. Similarly, for a step reduction, the integrator gives an output that falls linearly with time. Limits are provided to set the maximum acceleration and retardation values independently.

The curves shown in *Figure* 5.3 relate to a winding cycle when the driver's speed-control lever was moved to the full-speed position, the brakes were released after 3 s and the winder then accelerated under the control of the ramp generator. The winder was then given a short run at full speed and the driver's control lever was then returned to neutral, thus retarding the winder at the rate set by the ramp generator.

With the scheme the speed at any instant is determined by the output voltage from a tachogenerator being compared with the variable reference voltage to give an error voltage that ensures the required continuous control action. The speed curve in *Figure* 5.3 shows that acceleration and retardation are perfectly smooth with no sudden changes and the armature current is quite steady.

If the driver moved the control lever to the full-speed position, for example, without releasing the brakes, the output voltage from the ramp generator would rise at the rate corresponding to normal acceleration. If after a lapse of time the brakes were released, the speed error would be very large so that the winder would accelerate rapidly at a rate depending on the conveyance load until the speed corresponding to that called for by the ramp generator voltage was obtained. To avoid this circumstance an error limit feature in the control scheme prevents the ramp generator from integrating if the error rises above a predetermined value.

Speed-measuring Devices

The devices required to sense and measure deviations from desired values of the controlled quantities, and provide control signals to regulate the drive motor, are determined by the function of the driven unit.

If the function of the driven unit is controlled simply by regulating the speed of the motor, control is exercised by an electrical signal representing motor speed. For many drives this signal is obtained from some form of tachogenerator. The simplest form is a d.c. generator with a permanent-magnet field giving a d.c. voltage that is strictly proportional to speed. There are also a.c. tachogenerators giving an a.c. voltage which is rectified to provide a d.c. voltage. These generator types are coupled rigidly to the motor shaft (or to a shaft of the driven unit) and to ensure continuing accuracy special attention must be given to the mounting.

An alternative method of measuring speed is to use a toothed-wheel pulse generator. With this a toothed wheel is mounted on the shaft and rotates adjacent to a sensing device. As the projecting teeth pass the sensor, pulses are generated at a rate proportional to speed. Using an electronic unit to count pulses per unit of time, a speed signal can be obtained (see page 74).

Speed Control by Back e.m.f. Measurement

For some d.c. shunt-motor drives, a sufficiently accurate measurement of speed can be made in terms of the back e.m.f. (also referred to as counter e.m.f. or simply e.m.f.). With a constant field flux the back e.m.f. is approximately proportional to speed only if the load does not vary substantially so that the armature IR voltage drop is fairly constant. To enable the back e.m.f. to be used as a reasonably accurate speed signal under varying load conditions and over a required speed range, it is necessary to provide a measuring system which takes into account the IR drop.

The effect of IR drop is such that, unless compensation is provided by the control scheme, the motor cannot be operated at maximum load over the lower of a range of speeds. For example, with increasing armature current and a fixed armature applied voltage, the motor back e.m.f. falls to offset the increased IR drop, i.e. the motor speed falls in accordance with the drooping torque/speed characteristic of the shunt motor. Assuming that at full speed the IR drop is 10% of the motor back e.m.f., the inherent motor speed regulation, no-load to full-load, is also 10%. If then the required speed range is 10:1, and the control is set for 10% of full-load speed at no-load, with approximately 10% of maximum armature voltage and full-load, the motor will stall because to obtain full-load armature current the back e.m.f. must be zero so that speed must fall to zero.

To provide IR compensation, a signal proportional to armature current can be subtracted from the voltage feedback or added to the reference voltage. This follows from the basic equation for motor back e.m.f.

$$E = V_a - I_a R_a$$

where V_a is the applied armature voltage, I_a the armature current and R_a is the armature resistance. There are several methods of obtaining this signal.

Back e.m.f. Measuring Schemes

One method of measuring the back e.m.f. is by direct measurement of the armature current. Although an instrument shunt is used for ammeters, it does not provide a sufficient voltage for a control signal unless it is amplified. Some schemes use the back e.m.f. bridge shown in *Figure 5.4*. It consists of a resistor R in series with the armature and two resistors R_1, R_2 forming a voltage divider to which the voltage V_a is applied. With this scheme there is an output voltage V_o for use as a feedback signal proportional to back e.m.f. E. To avoid the power loss in the resistor R, a compole or series field winding is used to provide the resistance required.

A d.c. current transformer (a form of magnetic amplifier) can be used to provide an isolated current signal, as a voltage proportional to current, to give IR compensation in either the reference or feedback voltage.

With certain rectifier-fed d.c. motor drives, the alternating input current is proportional to the direct output current over the output-voltage range. An a.c. current transformer, designed to operate with a high-resistance load, can therefore be used to give an alternating voltage proportional to current which is rectified to provide a direct-voltage compensation signal.

Speed Control by Pulse-generator Frequency

To achieve a long-term speed-control accuracy of the order of 0·02% of top speed, a digital frequency-lock scheme can be used in conjunction with an analogue voltage scheme. With this, in addition to the voltage signal from a tachogenerator, the speed is also measured in terms of the frequency of the output of a pulse generator. This output is fed into a differential counter where its frequency is compared with a reference frequency directly

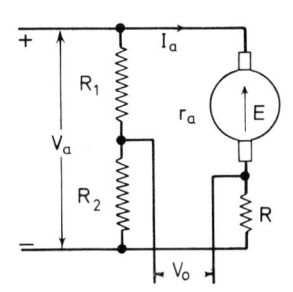

Figure 5.4. Back e.m.f. bridge

proportional to the desired speed. The counter and an associated amplifier produce a d.c. voltage output proportional to the time integral of the frequency difference, the output being positive or negative according to whether the pulse-generator frequency is lower or higher than the reference frequency. The resultant error signal provides a fine speed trimming signal.

Controlled Quantity Measuring Devices

With certain drives the speed of the motor is regulated in accordance with a desired value of a controlled quantity other than rotational speed. For example, the function of a pump drive may be to maintain a desired value of fluid flow through a pipeline or of a level in a tank. In such cases, the magnitude of the flow can be measured by an electronic flowmeter, and that of the level by a float-operated potentiometer, both of which provide a voltage signal for comparison with the desired value reference.

Similarly, if the ultimate function of the driven unit is to maintain a desired value of process temperature, then the control signal will be derived from sensors such as thermocouples, resistance thermometers or pyrometers giving an electrical output signal.

Some sensors do not provide an electrical output signal directly but this is obtained by a transducer that converts a physical or mechanical parameter into an analogue voltage. For example, the mechanism of a pressure gauge can operate the slider of a potentiometer giving an output voltage proportional to pressure.

Tension Control Schemes

The control schemes for strip material processing lines usually include devices and circuits for maintaining tensions required throughout the particular sections. Tension is applied by stretching the strip, and this is effected by regulating the speeds of the drives so that the linear speed at one is higher than the linear speed at the preceding drive.

When the strip is moving straight through the section, e.g. between two bridles as shown in *Figure* 5.5, tension is controlled by regulating the speeds of the two drives to maintain the appropriate linear speed differential, regardless of changing conditions causing disproportionate variations in the speeds of the individual drives. The speeds of the drives also have to be

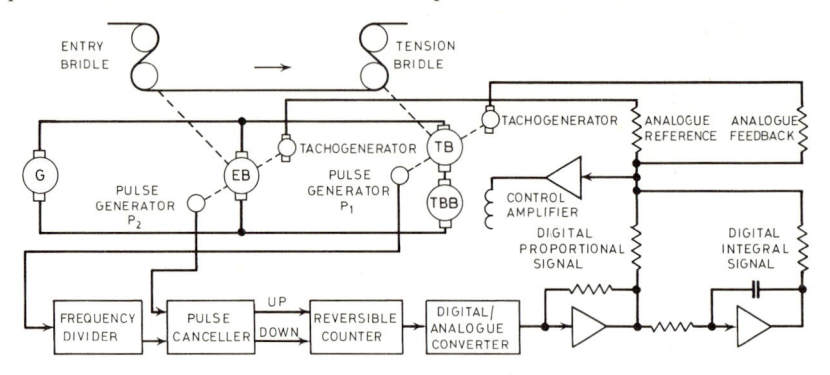

Figure 5.5. Precision speed-control scheme for bridle

regulated in accordance with the linear speed of the strip through the section of the process where one drive is regulated to establish the reference speed for the line, i.e. the line speed.

Precision speed control is essential for some types of process line where a critical value of tension is maintained by regulating inter-section speed ratios. A typical example is the control of the inter-bridle speed ratio on an aluminium tension levelling line where accuracies of 1 in 10 000 are required under certain conditions. To meet this requirement, digital counting techniques are used in conjunction with an analogue speed-control system.

Figure 5.5 shows a basic scheme engineered by AEI for the Rogerstone works of Alcan Industries Ltd. The entry and tension bridle motors EB and TB are supplied from generator G and the inter-section speed ratio regulated by the booster generator TBB. The analogue speed reference for TB is fed from a tachogenerator driven by EB and the analogue feedback is obtained from a tachogenerator driven by TB itself. This basic analogue system gives an accuracy of better than 1 in 1000.

Pulse generators P_1 and P_2 are driven by TB and EB respectively to provide the digital feedback signals for the counting equipment. A coincidence and alternate pulse canceller receives the pulses from P_2 directly and those from P_1 through a frequency divider. The average rate of entry of pulses on each channel can thus be made the same whatever speed ratio is selected. The bridle TB runs faster than EB to apply the tension required.

Any deviation in speed ratio from that originally set up on the frequency divider will cause differing rates of pulses to enter the pulse canceller. This causes pulses to be added to or subtracted from an 8-bit reversible counter, which is set to its middle position at start-up. A digital/analogue converter, biased to give zero output when the 8-bit counter is in the middle position, then produces an output voltage via its succeeding amplifier whenever a disturbance occurs. The output, with a polarity depending on whether the speed ratio has risen or fallen, is added to the analogue signal mixing point at the entry to the TBB control amplifier. This amplifier controls the excitation of the field winding of the booster generator TBB and therefore regulates the TB motor speed relative to that of the EB motor.

Strip-coiling Tension Control

When the strip is being coiled by a surface reeler, tension is controlled simply by regulating the speed of the reeler motor relative to that of the preceding drive to maintain a constant linear speed.

With a centre-spindle reeler, as the coil radius builds up the motor speed must fall to maintain a constant linear strip speed. This requirement is met in many cases by using a d.c. shunt motor with separately regulated supplies to the armature and the field winding. The speed is reduced as the coil builds up by increasing the field flux which also meets the requirement for the motor torque to increase as the retarding torque increases. Opposing the rotation of the armature, the retarding torque increases with coil build-up since the tension applied to the strip is acting at an increasingly greater radius.

The basic control requirement is that with constant line speed the tension and linear speed of the strip being coiled should also remain constant. With coil build-up the linear speed is maintained constant by reducing the rotational speed and since this is achieved by increasing the field flux, linear speed can be controlled in terms of the motor back e.m.f., which is proportional to rotational speed times field flux. At a given rotational speed, and therefore back e.m.f., the armature current is proportional to strip tension, so that in relation to any rotational speed providing the required linear speed, tension can be controlled in terms of armature current.

The torque/speed relationship is such that when coiling strip material at constant linear speed and tension, the centre-spindle reeler motor must develop constant power throughout the coil build-up. Although the motor torque increases proportionally with coil radius, and its rotational speed falls in inverse proportion to coil radius, the power, i.e. torque times speed, remains constant.

Strip Coiler Control Scheme

Figure 5.6 shows an AEI control scheme for a d.c. motor using a current regulator and a back e.m.f. regulator. The current regulator controls the motor armature current by varying the armature supply voltage, while the back e.m.f. regulator controls the motor field supply voltage and hence the flux.

The reference for the current regulator is made up of two separate signals, the first being the tension reference set by the operator, which is in fact the current reference when the line is running at a steady speed. During speed changes, however, as additional torque is required for acceleration, a second reference signal is added to the tension reference. This is the inertia compensation signal which is dependent on the strip speed change rate, the coil diameter and the strip width.

The function of the back e.m.f. regulator is, in effect, to keep the motor flux in step with the coil diameter. A signal proportional to strip speed is compared with one proportional to back e.m.f. in an amplifier that produces an output when the back e.m.f. tends to fall as the coil builds up and the retarding torque increases. The amplifier output drives a follow-up device which continuously strengthens the motor field to maintain the balance of back e.m.f. and speed. Since the coil diameter builds up slowly the follow-up device need not be very fast and a motor-operated rheostat is often used.

If the strip breaks, the current regulator tends to maintain the armature current and the motor starts to accelerate rapidly causing the motor back e.m.f. to rise. This reverses the output of the e.m.f. amplifier, which is switched by diodes from the follow-up device to the current amplifier to

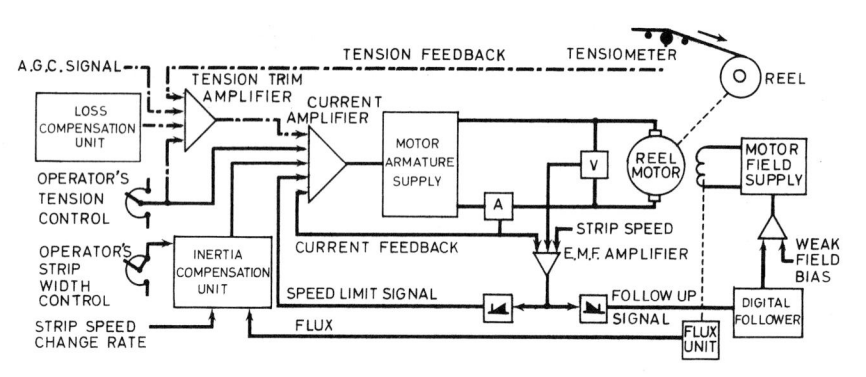

Figure 5.6. Coiler strip tension control scheme

balance the current reference and hold the reel at a peripheral speed slightly higher than the strip speed.

The same control is also used when threading the mill on the fly, that is running at thread speed. As the strip approaches the reeler, the tension reference is switched on, causing the reeler to run slightly above thread speed. On reaching the reeler, the strip is wrapped around it by the belt wrapper, and as soon as the strip has been gripped the reeler speed falls to thread speed. The reeler is thus brought out of the speed limit condition, and tension is established smoothly.

Where the strip tension must be controlled very accurately or where a wide tension range is required, a loss compensation signal is provided to correct the current reference for the mechanical losses of the drive. When automatic gauge control is installed, a further signal is provided to modify the tension reference as required by the control. All these functions are indicated in *Figure* 5.6 by broken lines.

The disadvantage of the scheme is that the reeler accelerates on weak field at the start of a coil and consequently limits the acceleration and stopping rates of the process line. An alternative scheme allows the motor to accelerate first on full field from zero to base speed and thereafter by field weakening. The current regulator is replaced by a tension regulator, and the motor flux is controlled by a voltage spillover unit. The torque must be adjusted continuously to compensate for the variations in the losses with changing speed. To simplify the method of compensating for the losses, the motor speed/loss relationship should be reasonably linear. To this end the

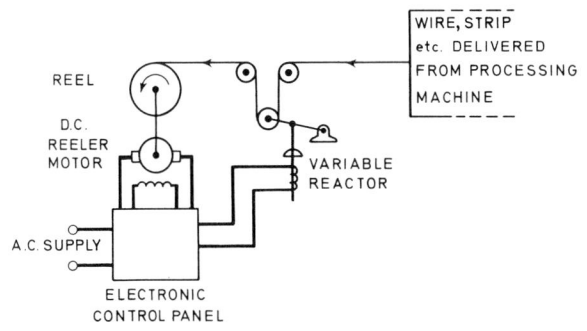

Figure 5.7. Tension control by dancing roll-operated variable reactor

motor is often cooled by a separate motor-driven fan unit instead of a shaft-mounted fan. This simplifies the design of the control scheme because the motor mechanical losses do not include the power absorbed in driving a fan. Also, with a separate fan unit the constant cooling air supply minimises temperature variations and consequently changes in the field strength of the reeler motor.

Reeler Speed Control by Tension Measurement

Instead of providing loss compensation, a tensiometer can be used to measure tension and give a signal which, compared with the tension reference value, continuously regulates armature current and torque (*Figure* 5.6).

An analogue voltage representing the actual tension in the material can be used as a feedback signal for direct tension control. This ensures accurate control, particularly with light tensions when the mechanical losses in the reeler and the drive are of the same order as, or are greater than, the power transmitted to the material being wound.

Tension feedback signals are obtained from devices such as inductive transducers coupled to a dancing (or floating) roll, load cells and tensiometers.

With the tension control scheme shown in *Figure* 5.7 a dancing roll floating on a loop of the material is coupled to the core of a variable reactor. Any variation in the loop causes the roll to move up or down to adjust the core and alter the inductance of the reactor and therefore the value of the output signal regulating the motor speed. With a centre-spindle reeler, as the coil radius increases, the roll tends to rise and change the signal output to reduce the speed of the drive motor. The scheme is also used to maintain constant tension in material moving through a process since any tendency for the linear speed to rise or fall causes an upward or downward movement of the roll and a fall or rise in the rotational speed. Dancing roll-operated potentiometers are also used to provide control signals for processing lines, e.g. where loops of material must be maintained at specific lengths (see page 126).

A load cell converts minute linear movements into electrical signals proportional to the mechanical pressure applied.

A tension feedback signal is obtained from a load cell in the scheme shown in *Figure* 5.8, devised by Laurence, Scott & Electromotors, for a Ward-Leonard reeler drive and including a tension tapering facility. The output

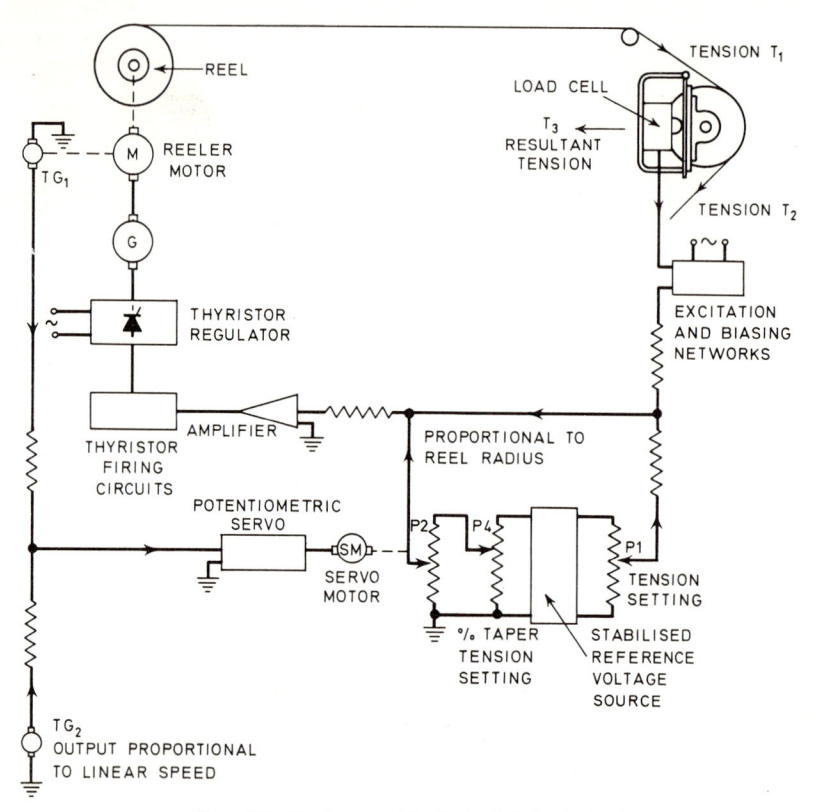

Figure 5.8. Tension control by load cell feedback signal

voltage produced by the load cell due to the weight of the roller is neutralised by a biasing network so that its effective output voltage is proportional to the tension in the material. This output voltage is compared with a voltage proportional to the desired tension as set on potentiometer P_1 and the error voltage is fed into the control system amplifier. The output from the amplifier controls the thyristor regulator which adjusts the generator voltage and therefore the torque of the reeler motor M, which is determined continuously by the comparison of the load cell voltage (the measured value) and the P_1 voltage (the desired value).

To obtain well-wound rolls of paper and some types of plastics films it is usually necessary to reduce reeling tensions gradually during the build-up of the roll. The amount of taper or reduction in tension required may vary with the thickness of the material so the control scheme includes a facility for tapering the tension by an adjustable predetermined amount.

With a tension feedback scheme this facility can be provided by continuously computing roll radius and automatically adjusting the desired tension signal progressively to obtain the appropriate rate of reduction in tension. Roll radius can be computed by dividing the voltage proportional to the linear speed of the material (V_1) by the voltage proportional to the rotational speed of the motor (V_r). This follows from the equation

$$\text{linear speed (m/min)} = \text{rotational speed (rev/min)} \times 2\pi \text{radius } (R)$$

therefore

$$2\pi R = \frac{\text{m/min}}{\text{rev/min}}$$

and

$$R = \frac{V_1}{V_r}$$

Using voltages from the rotating-speed tachogenerator TG_1 and the linear-speed tachogenerator TG_2, the division is done by a potentiometric servo giving an output to the servomotor SM driving the potentiometer P_2. This feeds a continuously variable signal, proportional to roll radius, into the summing network, which also receives a tension control signal from the load cell. The combined effect of these two signals is to reduce the reeler motor speed at such a rate that the linear speed of the material and therefore the tension applied are also reduced as the roll builds up. Potentiometer P_4 in *Figure* 5.8 is used to adjust the voltage applied to potentiometer P_2 and therefore the percentage of taper tension.

Loop Position Control Schemes

When strip material is moving continuously through a processing line, one or more free-hanging loops may be used for tension relieving or to serve as buffer stores to allow for temporary changes in line speed. In this case the speeds of the relevant drive motors are regulated to keep the loop positioned within a controlled zone. Various schemes have been developed to enable changes in loop length to be sensed by a system giving an electrical output signal initiating regulation of motor speeds as required.

A photoelectric control scheme is an effective method of maintaining the position of a free-hanging loop within a controlled zone. Such a system must not be affected by ambient light, by aging of the light source or photocell, or by the accumulation of a certain amount of dust on the optical system. *Figure* 5.9 shows an AEI scheme meeting these requirements and providing for the automatic correction in the appropriate sense of any deviation from the centre of the zone.

A fluorescent lamp is located vertically in the pit and its centre is level with the normal position of the bottom of the loop. A window with alternate clear and dark bands is fitted in front of the lamp. A detector unit has a motor-driven drum carrying four lenses at 90° spacing. Each lens scans the length of the lamp unit in turn and causes the image of the unit to traverse a slit. Behind this slit is a photocell and its output is fed into a transistor amplifier, which produces a succession of shaped pulses, each pulse corresponding to a clear band on the light unit. The number of pulses counted during each

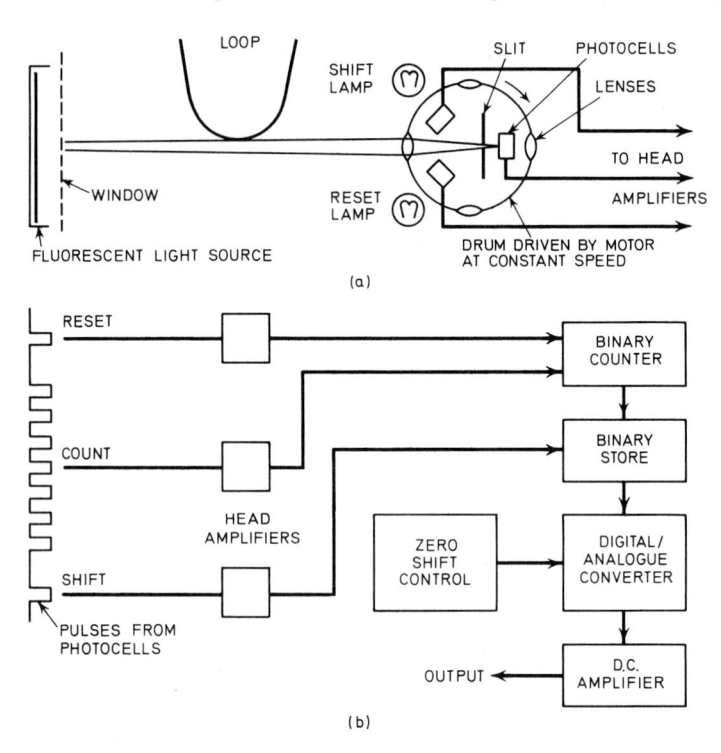

Figure 5.9. Control of free-hanging loop

scan depends on the position of the loop, and the count is therefore proportional to the height of the loop above the control zone.

Each train of pulses is gated on and off by signals from small 'reset' and 'shift' lamps into photocells mounted near the edge of the drum. The first gating pulse resets the binary counter to zero and, after the train of impulses has been counted, the second gating pulse transfers the count to a binary store which is then up-dated at the end of each scan. The binary store feeds into a digital to analogue converter, the output of which is given a zero bias before being fed to the d.c. input amplifier of the appropriate speed regulator. The length of the strip in the looping pit is thus adjusted to the desired value.

In metal rolling mills a loop of material is often introduced between a pair of stands to ensure freedom from tension despite variations in the temperature of the rolled stock. The various loop control schemes used all have a means of forming the loop initially, a scanning system detecting any deviation of the loop from the reference position and a means of translating the deviation into a position error. This error is usually fed to the speed control scheme for the driving motor of the downstream stand of the pair in order to correct its speed and so restore the loop to its reference position. The error can be fed to the upstream stand if desired.

During the threading of single-strand rod mills, the rod is usually allowed to form a catenary. In narrow strip rolling an upward loop is sometimes used, the necessary slack being produced by the impact speed drop or it is introduced deliberately when setting up the speeds of two stands.

Figure 5.10 shows an AEI scheme for producing and maintaining an upward loop. A persuader roll rises to deflect the strip into the required loop when a hot-metal detector indicates that the metal has emerged from the downstream stand. A second hot-metal detector at the entry to the upstream

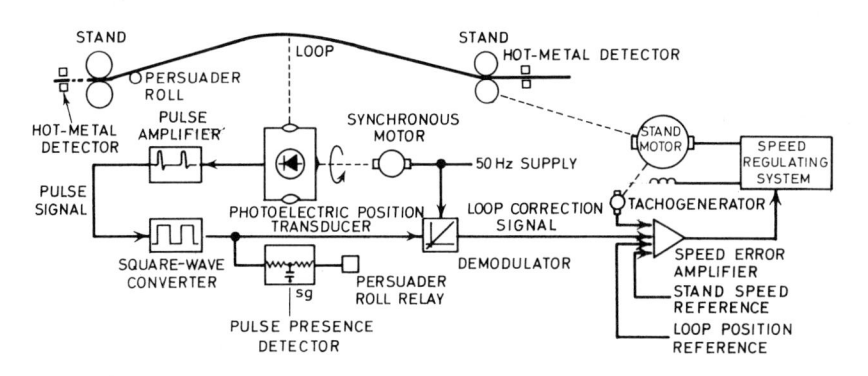

Figure 5.10. Control of upward loop. (The strip is moving from left to right)

stand is used to detect the disappearance of the tail end and to lower the persuader roll in preparation for the next strip.

To regulate its position the loop is scanned by a position transducer comprising a synchronous-motor-driven rotating head with an optical system which focuses an image of the hot metal on to a photocell sensitive to infrared radiation. The synchronous speed and optical system are so arranged that the infrared cell receives a stimulus 100 times per second. These short duration pulses are amplified and fed into a bistable (flip-flop) circuit to convert them into a square-wave signal, the leading edges of which correspond to the leading edges of the pulses. The square wave is then fed through a demodulator circuit, which uses the 50 Hz supply voltage as the gate reference. The phase angle of the square wave compared with the gate reference is a measure of the delay in the original firing pulse. Thus, with the scanning head aligned so that the mean loop position gives zero output from the demodulator, any deviation in loop position will produce a demodulator output voltage proportional in magnitude to the degree of displacement and polarised to indicate the sense of the displacement. This output is applied to the speed reference of the downstream motor to modify its speed and so restore the loop to the reference position.

If there is a series of loops interspersed between the stands, only the first persuader roll requires a hot-metal detector signal to initiate lowering. Each subsequent persuader roll is lowered when the tail end disappears from the field of view of the scanner regulating the previous loop, as indicated by a loss of output from the pulse presence detector.

Control Amplifiers

A control scheme accepting low-level signal inputs includes one or more control amplifiers with associated devices and circuits to process and mix the signals from various sources. Typical amplifier networks for performing different control functions are shown in this and other chapters.

There are four main types of control amplifiers: thermionic valve; rotating; magnetic; transistor. In general, thermionic-valve types have now been superseded by transistor types.

Rotating amplifiers, e.g. amplidynes, are still in service for schemes engineered in past years, but the subsequent development of static magnetic amplifiers resulted in these being used when practicable instead of rotating types.

Magnetic Amplifiers

Similar in principle to the transductor (see page 65), the magnetic amplifier can provide directly an output for a shunt field winding regulated by various incoming signals to integral control windings. *Figure* 5.11 shows a control scheme used by English Electric for a d.c. shunt motor drive for a billet mill stand, with a magnetic amplifier supplying the motor field. The speed reference is set up by the operator adjusting the coarse and vernier rheostats CR and VR which are in circuit with the speed control winding S and the speed reference exciter RE driven by the motor M.

A special type of magnetic amplifier developed some years ago, the flux-reset magnastat, is energised by a 1000 Hz supply from a transistor oscillator to ensure a fast response to signal inputs. As shown in *Figure* 5.12, the basic 50 Hz full-wave amplifier includes a saturable reactor and an arrangement of rectifier diodes to provide a d.c. output to the load. The d.c. output is regulated by a variable voltage or a variable resistor connected to the control terminals. Alternatively, the control input may be derived from a transistor amplifier.

With the full-wave magnastat, a change in control made at the commencement of any half-cycle of the a.c. supply is fully effective in the output of the succeeding half-cycle. A change made during a half-cycle is not completely effective until the second successive half-cycle. Hence, a full-wave magnastat under optimum conditions will respond with a half-cycle of the a.c. supply. The early flux-reset amplifiers were energised at 50 Hz so that, with a maximum delay in the response of one cycle, the time was 0·02 s. With the higher supply frequency the time is much shorter.

Magnastats are designed with several control windings for signal mixing

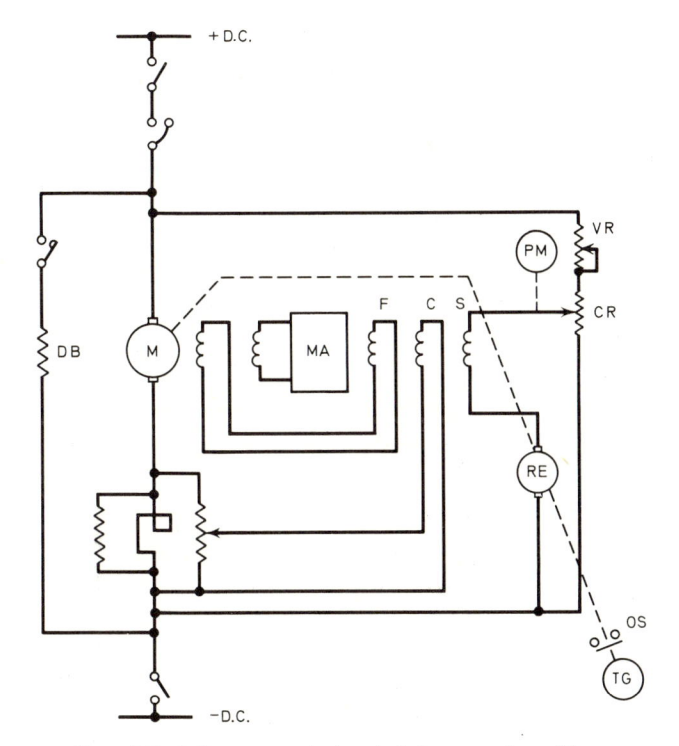

Figure 5.11. D.C. motor control scheme including magnetic amplifier

C = Compounding control winding
CR = Coarse speed setting rheostat
DB = Dynamic braking resistor
F = Feedback control winding
M = Stand motor
MA = Magnetic amplifier
OS = Overspeed
PM = Pilot motor
RE = Speed reference exciter
S = Speed control winding
TG = Tachogenerator
VR = Vernier speed setting rheostat

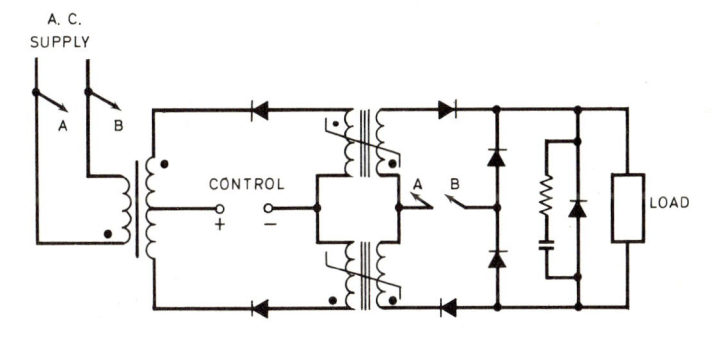

Figure 5.12. Flux-reset magnastat amplifier

and for d.c. outputs up to 30 kW at 350 V to function as shunt-field current regulators for large motors (see page 121).

Transistor Amplifiers

Recently developed motor speed control schemes usually incorporate transistor amplifiers. *Figure* 5.13 shows a scheme including a transistor amplifier and used for the control of the field excitation of a Ward-Leonard generator. The field is supplied by a thyristor unit controlled by a firing circuit incorporating a saturable reactor (see page 65) and requiring a current output from the control amplifier. The control winding of the saturable reactor is in series with the output of the last amplifier stage Tr_4. In addition

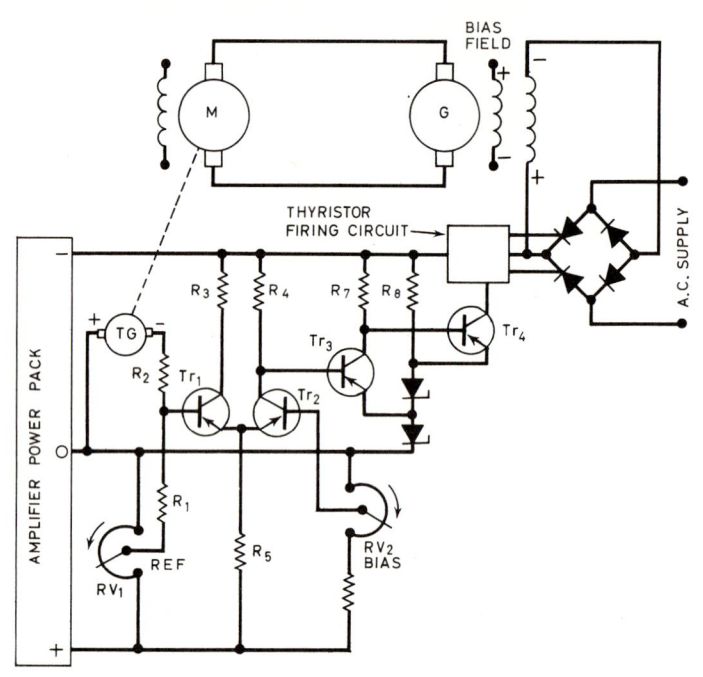

Figure 5.13. Ward-Leonard generator field excitation scheme including transistor control amplifier and thyristor regulator for field power supply

to the main field the generator has a bias field which is excited to oppose the main field for the purpose of neutralising the residual flux to obtain precise regulation of the motor at the lowest speeds (see page 54).

The speed reference for the scheme is provided by the voltage set by the potentiometer RV_1 and obtained from the d.c. power pack. The difference between the reference and tachogenerator voltages (the error voltage) is applied to the base of transistor Tr_1 which with Tr_2 forms a so-called long-tailed pair with R_5 as the common-emitter resistor and R_3 and R_4 as collector resistors. The base of Tr_2 is connected to the slider of potentiometer RV_2 so that the output of the amplifier may be adjusted to zero for zero deviation input. The long-tailed pair is followed by the two amplifier stages Tr_3 and Tr_4.

When RV_1 is set to increase speed, the base of Tr_1 is made more positive thereby reducing its collector current and increasing the collector current through Tr_2. The base voltage of Tr_3 thus becomes less negative, its collector current decreases and increases the negative base voltage on Tr_4. The current through the control winding of the firing unit is increased to advance the firing angle of the thyristors thus increasing the input to the generator field. Setting RV_1 to reduce speed produces current and voltage changes in the opposite sense and with deviations in the error voltage due to load fluctuations, constant speed is maintained by automatic variation of the current and voltage values determining the excitation of the generator field.

Control-amplifier Signal Circuits

In a complex scheme with several signal inputs, in effect the control amplifier functions to compute an output signal which will initiate a control action necessary to maintain the steady state of a drive system or change the state without any tendency towards instability, e.g. a sustained variation of the error-voltage signal above and below a particular value which at any instant is the desired value. To avoid instability a stabilising *RC* (resistance-capacitance) feedback circuit connected from the output to the input of the amplifier serves to control the error signal so that it initiates control action at a rate that ensures rapid but smooth regulation of the drive system. Other input signals are used to limit current, acceleration and deceleration while optimising the response time of the system under all conditions of operation.

Reference-voltage inputs may be adjusted by biasing circuits to meet changing control requirements, e.g. to change from armature-voltage to field-weakening speed regulation of a d.c. motor (see page 110). Shaping circuits are used to obtain a control signal required to vary with time in accordance with a curve of a particular shape. Since reference-voltage values must be maintained constant to ensure precise control of the drive, the voltages are obtained from stabilised supply units.

Combinations of different types of amplifiers may be used for particular schemes, but the emphasis today is on the use of solid-state devices integrated with control circuitry to form modules for different functions. Standard ranges of such units are available for engineering into schemes for various control requirements. Alternatively, assemblies of components are available in the form of controllers designed for various types of drive. For rectifier-fed d.c. drives there are ranges of standard units incorporating all the control and power-supply components providing the facilities for different control requirements.

In addition to the control and power regulation components of an automatic drive scheme, standard types of a.c. contactors and circuit-breakers, with associated protective devices, are necessary for switching the incoming supply to the power regulating equipment. In some cases motor starters may also be required so that the scheme may involve automatic control of the a.c. supply switching and current regulating gear.

Chapter 6

A.C. Motor Drives

For the simplest on/off schemes requiring only the direct-on-line switching of a squirrel-cage motor, the basic control circuits are similar to those shown in *Figure* 5.1 for starting or stopping the motor by the operation of an on/off device such as a thermostat, pressure switch or float switch.

In general, on/off motor control schemes are used only when fairly wide variations in the controlled quantity are acceptable. For example, the level of the fluid contents of a storage tank can often be allowed to fall considerably before the replenishing pump motor is started.

Single-motor Level-control Schemes

For this type of pump drive motor starting may be initiated by a control signal derived from electrodes located in the tank. The arrangement is shown in *Figure* 6.1 illustrating the application of a Fielden Electronics *Tektor TT*6 level controller. The signal input to the control unit results from the variation of the capacitance of the system with changes in the level of the contents—a liquid, powder or granular solid. The two electrodes are normally inserted into the side of the container at heights where control action is required. Alternatively, the electrodes may be mounted vertically with their tips reaching just beyond the required control levels.

With the Fielden *Noflote* level-control system, used for conductive liquids, the electrodes are arranged vertically and the control signals result from the variation with level of the resistance of the system. One electrode is longer than the other and the lengths are such that the tips correspond to the required control levels. When the liquid rises and touches the high electrode there is a low-resistance path between the two electrodes which initiates a signal for switching off the pump motor which has been running to fill the tank. The motor is not switched on again until the level falls below the tip of the low electrode so that there is a high-resistance path between it and earth.

Valve Actuator Drives

Many types of fluid-flow regulating valves are operated by actuators powered by 3-phase squirrel-cage motors that are switched by an on/off scheme in response to signals from process controllers. With the scheme shown in *Figure* 6.2 one signal from the process controller is fed to a **KPE** Controls *Kom-Pac* 1 set-point amplifier and another signal from a feedback potentiometer in the actuator is fed to a second amplifier. The two signals are compared continuously and are practically equal until the process-controller signal changes to initiate an adjustment of the valve and the signal differential results in an output from the appropriate amplifier to energise one of two relays used to switch the actuator drive motor in either the forward or reverse direction. As the valve is adjusted the feedback potentiometer slider moves to equalise the signals. The valve movement is proportional to the input signal.

In order to prevent the motor hunting in either the forward or reverse directions, a dead-band adjustment of up to 20% of the input span is provided on each amplifier. The dead band ensures that the signal differential

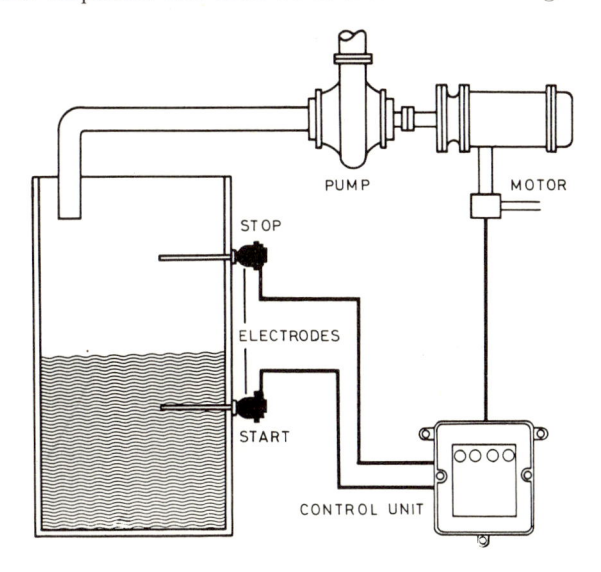

Figure 6.1. Level-control scheme

must exceed a particular value before valve adjustment is initiated. It is set by a potentiometer according to the transient fluctuations likely to occur during the operation of the plant.

For the control of actuators of valves subject to more or less continuous adjustment in small increments, static switching may be used. The basic *Ascron* thyristor switching system shown in *Figure* 3.6 for the control of induction regulator drive motors is also used for the *Limonic* standard control system for actuators produced by Limitorque Valve Controls for operating valves and dampers employed for a variety of modulating duties.

Sequential Motor-starting Schemes

A control requirement with induction-motor drives for continuous processing or materials handling plants is for starting the motors of various sections in a

88   A.C. Motor Drives

predetermined sequence. For this purpose an electrical interlocking scheme ensures that one contactor starter is closed before the next one in the sequence can be closed.

Figure 6.3 shows a simple interlocking scheme applied to three motors each driving a separate conveyer forming part of a continuous materials handling system. The sequence interlocking is arranged so that contactor B cannot be closed until contactor A has closed, and contactor C cannot be closed before B. Thus the starting sequence is A-B-C and is opposite to the flow of material.

The scheme ensures that conveyer B must be running before conveyer C can feed material to it; and similarly conveyer A must be running before B

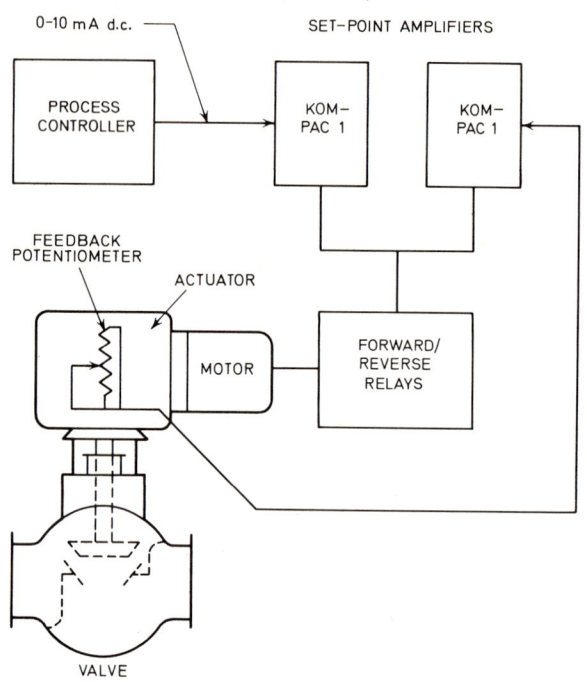

Figure 6.2. Valve actuator drive

can feed it with material. If conveyer B stops, then from the diagram it will be seen that conveyer C also stops and cannot feed material to the stationary conveyer B so that there is no risk of a build up. At the same time conveyer A can continue to run and discharge whatever material it is carrying.

The scheme depends on the use of interconnected auxiliary contacts and it should be noted that those on the isolating switch are duplicated. This is a safety precaution to ensure that when a circuit is isolated, there is no danger of feedback via interconnected wiring making some point in the unit alive when it appears to be completely isolated and available for a man to work on.

Interlocking schemes are usually provided with some means of enabling tests or inspection on one circuit to be carried out without the need to have the other circuits operating. In *Figure* 6.3, rotary switch (7) is provided for this purpose and it has two positions to allow for in-sequence or out-of-sequence switching.

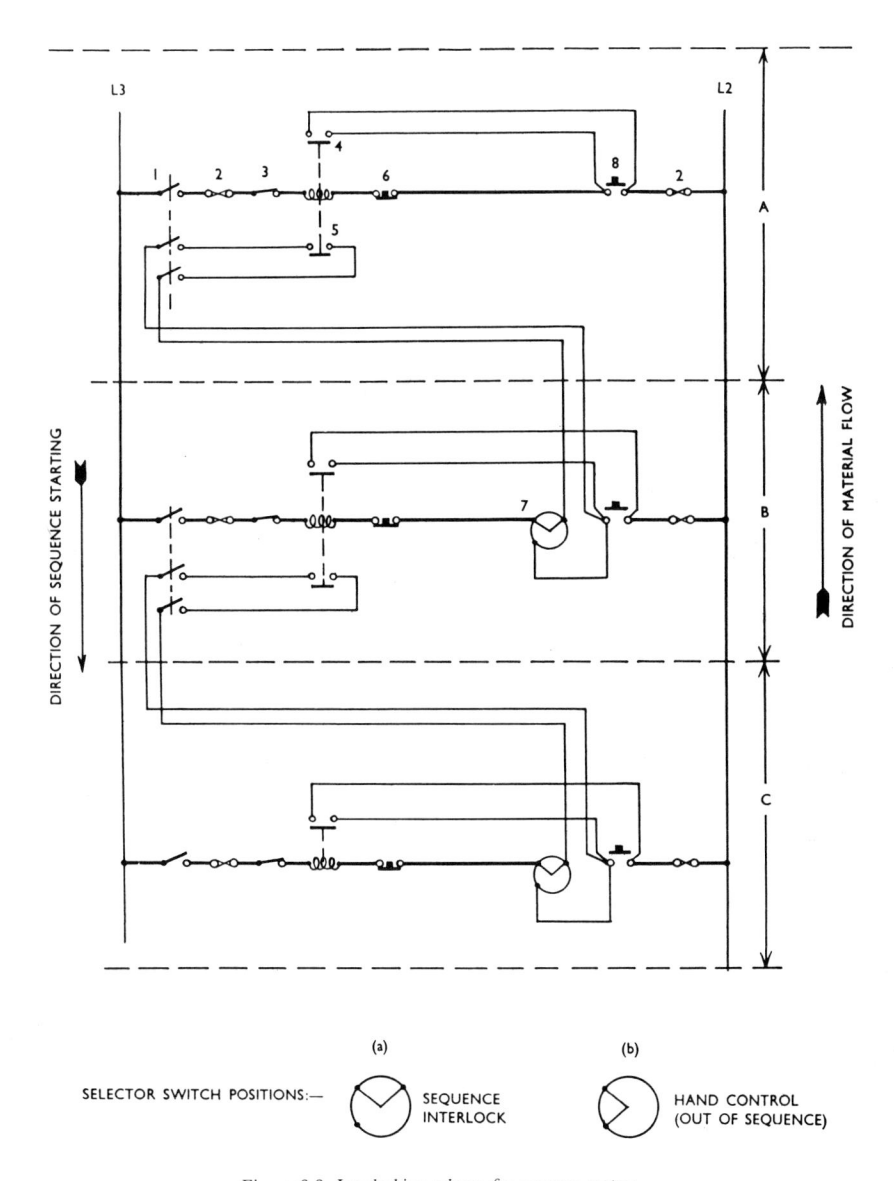

Figure 6.3. Interlocking scheme for conveyer motors
1. Aux. contacts on isolator (normally open)
2. Control circuit fuses
3. Overload trip switch
4. Retaining contact on contactor (normally open)
5. Interlock contact on contactor (normally open)
6. 'Stop' push button
7. Rotary selector switch
8. 'Start' push button
L2 and L3—Supply across two lines or from bus-wires

In other circumstances it may be necessary in certain operations to feed material direct from conveyer C to conveyer A, missing B. A modification of the rotary switch, or an additional bypass switch can be wired into the control circuit to provide this facility.

The scheme in *Figure* 6.3 requires that each contactor be closed in turn by an operator pressing the start pushbuttons (8) in the order A-B-C. To indicate that each contactor has actually closed, a red light is usually provided on each circuit.

In an alternative scheme, automatic sequence interlocking ensures that after closing contactor A, all succeeding contactors will close automatically. This type of scheme is shown in *Figure* 6.4. Contactor A is first closed by the operator using either the local or remote start button. Attached to this contactor is a time-delayed auxiliary contact (5). After the time delay, the contact closes to complete the control circuit for contactor B which initiates the closing of contact C in the same way, and so on to the end of the sequence.

Contactor A in *Figure* 6.4 can be operated either by local or remote control but B, C and D only have local controls. Rotary switches are provided in each circuit, that in A to select either local or remote control and those in B, C and D to select between in sequence or out of sequence. If the B, C and D rotary switches are set at out of sequence, all four motor starters can be operated individually and in any order.

The simple sequence-control scheme shown in *Figure* 6.5 could be applied to a commercial refrigeration plant using a fan and a pump, both driven by squirrel-cage motors with direct-on-line starters, and a compressor driven by a slipring motor with a stator-rotor starter. In the scheme for this plant automatic control is provided by a remote thermostat (15) and a pressure cut-out (14). Operation of the plant requires that both the fan and the pump must be running before the compressor can be started. The contactor (5) controlling the fan must therefore close before the contactor (6) controlling the pump, and only when (6) is closed can the contactor (7) be closed to start the compressor.

When running on automatic control (with all rotary switches set at auto), operation of the high-pressure cut-out to open its contact (14) shuts down the compressor, leaving both the fan and pump running. When the pressure falls, contact (14) recloses and the compressor re-starts. Operation of the thermostat to open its contact (15) will shut down the complete plant until a temperature change permits contact (15) to reclose and re-start the plant in the proper sequence. The rotary switch enables the plant to be started manually if required.

Although with schemes more complicated than those described the interlocking will be more involved, they will be generally similar in principle. Some schemes require that a complete group of motors must be started in sequence before another group. In this case two separate control boards must be interlocked.

Control Schemes for 3-phase Commutator Motors

The 3-phase commutator motor is applicable to many drives with a control requirement either for a constant speed or for an infinitely variable speed.

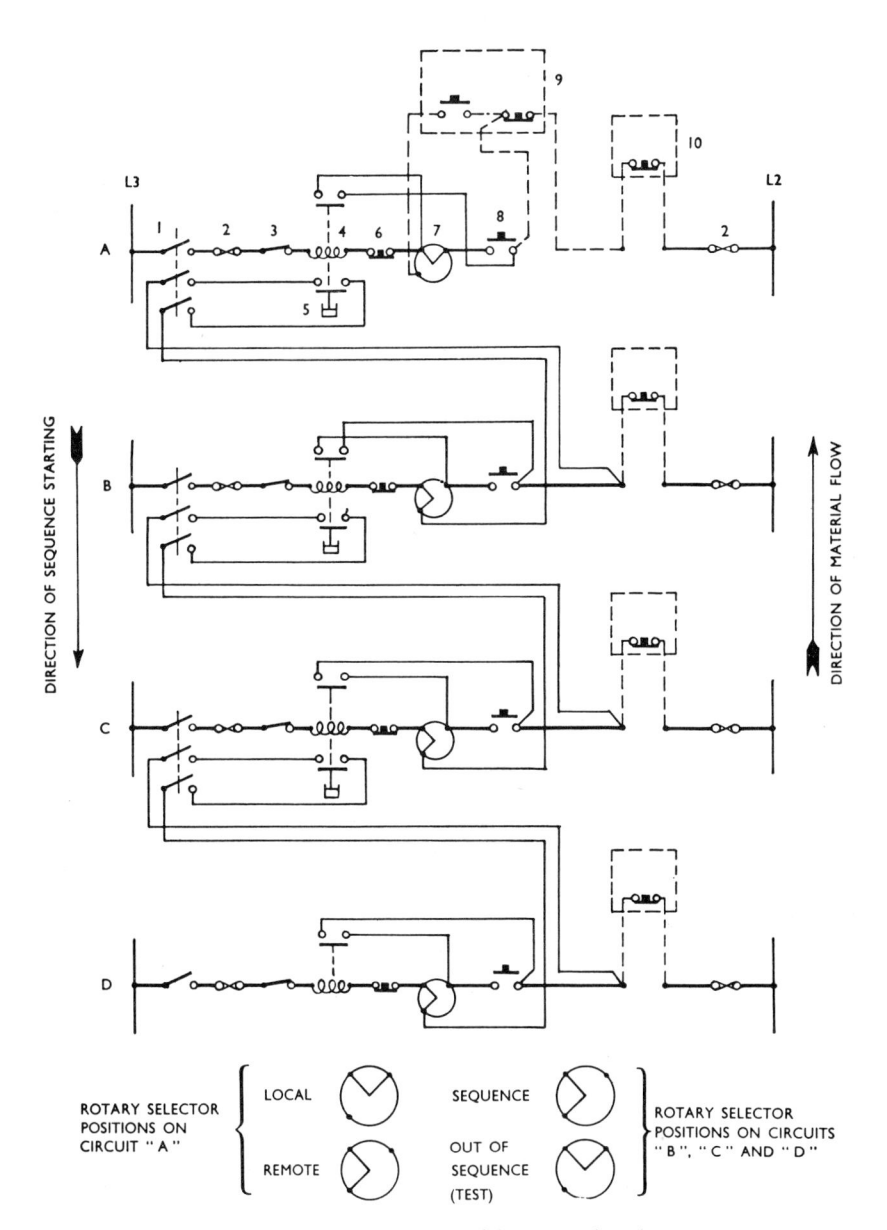

Figure 6.4. Automatic sequential motor starting scheme
1. *Aux. contacts on isolator (normally open)*
2. *Control circuit fuses*
3. *Overload trip switch (normally closed)*
4. *Retaining contact on contactor (normally open)*
5. *Time delayed contact on contactor (normally open)*
6. *'Stop' button*
7. *Rotary selector switch*
8. *'Start' button*
9. *Remote 'start/stop' station*
10. *Remote emergency 'stop' station*
L2 and L3—Supply across two lines or from bus-wires

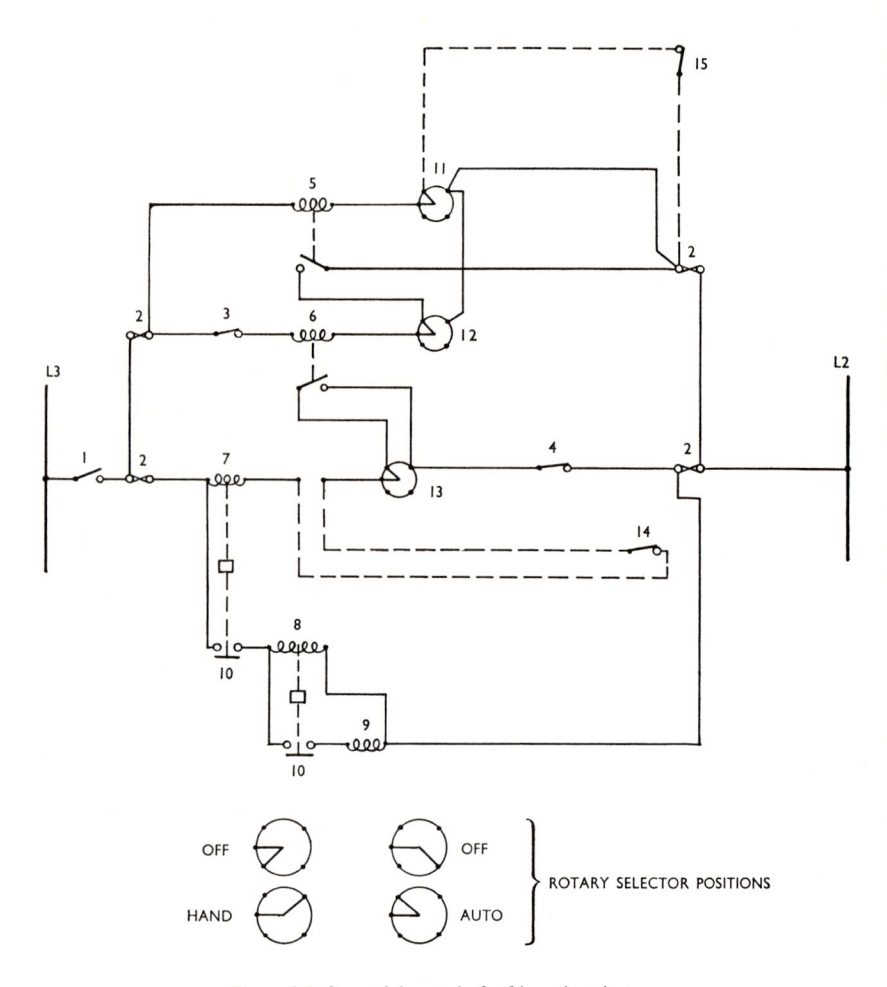

Figure 6.5. Sequential control of refrigeration plant
1. Aux. contact on isolator (normally open)
2. Control circuit fuses
3. Pump motor overload trip switch (normally closed)
4. Compressor motor overload trip switch (normally closed)
5. Fan motor contactor coil
6. Pump motor contactor coil
7. Compressor motor contactor coil (stator)
8. Compressor motor rotor contactor coil (No. 1)
9. Compressor motor rotor contactor coil (No. 2)
10. Delayed closing aux. contacts on 7 and 8
11. Rotary switch in fan circuit
12. Rotary switch in pump circuit
13. Rotary switch in compressor circuit
14. High pressure cut-out
15. Thermostat
L2 and L3—Supply across two lines or from bus-wires

A.C. Motor Drives

Automatic speed regulation of both stator-fed and Schrage types of motor can be effected by a scheme controlling the operation of a pilot (or servo) motor driving respectively an induction regulator or the brushgear rockers.

Stator-fed Motor Control Systems

For the speed regulation of the stator-fed type N-S 3-phase commutator motors, Laurence, Scott & Electromotors has developed three systems to meet different control requirements determined by the performance characteristics of the drive. *Figure* 6.6 shows the main components and circuits of the *Asrec* electro-mechanical relay scheme. *Figure* 6.7 is a more detailed diagram including the pilot motor control circuit.

The scheme operates from a d.c. supply provided by a transformer-rectifier unit. The desired speed is set by the control potentiometer which is

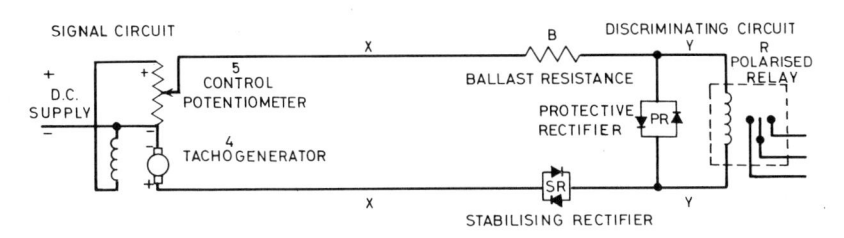

Figure 6.6. Basic Asrec *control system*

connected to the same d.c. supply as the field winding of the tachogenerator providing the measured speed value. With this arrangement, a stabilised voltage supply is not required as the tachogenerator and the control potentiometer voltages will together follow line-voltage fluctuations provided that the tachogenerator is unsaturated and the heating effects are small—which is the case in practice.

The control and tachogenerator voltages oppose one another and the error voltage is applied to the discriminating circuit. The main discriminating element is the polarised relay R. It consists of a permanent magnet and a spring-controlled moving coil carrying a contact which makes on either of two fixed contacts according to the polarity of the error voltage. The closure of the contacts initiates operation of either of two contactor circuits for energising the pilot motor to adjust the induction regulator in order to 'accelerate' or 'retard' the main motor.

The miniature polarised relay is very sensitive operating with an input of a few microwatts and giving a power amplification of about one million. To avoid the flutter experienced with some polarised relay schemes owing to small random variations of voltage, a dead-band effect is obtained by using a small two-way rectifier SR as a nonlinear resistor in series with the coil. It has the characteristic voltage/current curve shown in *Figure* 6.8, so that up to a particular applied voltage the current is virtually zero. Consequently, with small and/or random voltage fluctuations the relay is not deflected and there is no wear of moving parts and contacts. To protect the relay

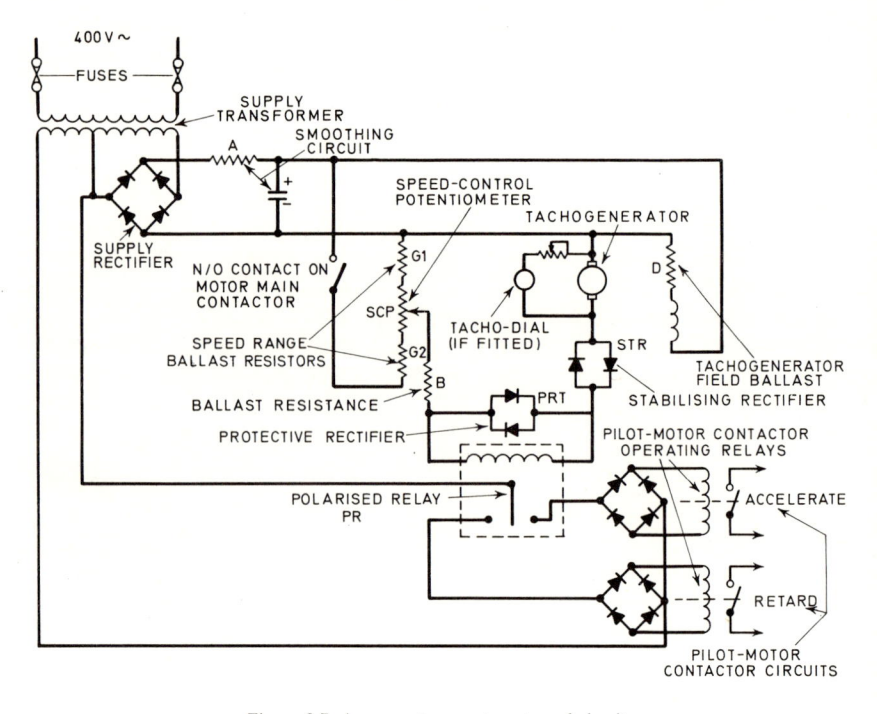

Figure 6.7. Asrec *system components and circuits*

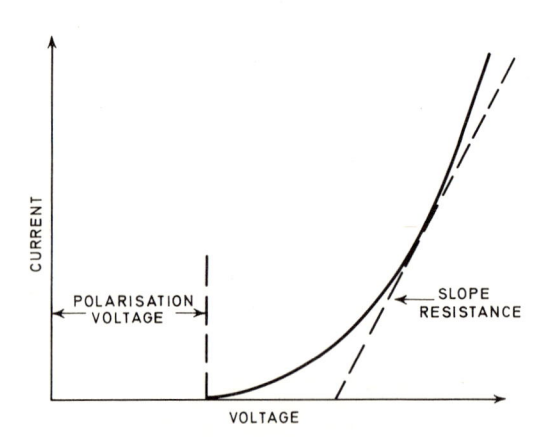

Figure 6.8. Typical voltage/current curve of nonlinear resistor (rectifier element)

against the effect of overvoltages there is the combination of the series ballast resistance B and the two-way shunt rectifier PR also serving as nonlinear resistor.

The scheme shown in *Figure* 6.7 includes two relays controlling the pilot-motor contactors (not shown) but the smaller pilot motors are switched directly by the relays.

Developed for the control of the induction regulator driving motor, the *Asrec* system has a relatively slow response compared with other systems developed for certain high performance drives, but it is completely adequate for drives which can be powered by type N-S motors.

N-S Motor Automatic Drives

The *Asrec* system can control type N-S motor drives to maintain a constant speed or to vary the speed continuously or intermittently as required.

One example of an *Asrec* constant-speed application is a calender drive requiring a preset speed to be maintained. At the same time a very wide range of preset calendering speeds is necessary together with automatically controlled low inching speeds.

To maintain a required constant speed the potentiometer SCP (*Figure* 6.7) is simply set to the desired value. When the speed of the driven unit has to be varied for a specific purpose, the control signal must also vary to adjust the motor speed as required. For instance, the requirement may be for the motor to run continuously and for its speed to be varied as necessary to maintain a controlled quantity within specified limits, e.g. a pump motor speed is varied to maintain a sufficient head of liquid in a storage tank to provide for changes in the rate of outflow.

Figure 6.9 shows the basic components of a variable-speed pump drive with a proportional level-control scheme which adjusts the motor speed between minimum and maximum values according to the outflow from the tank. A float-operated potentiometer functions as a transducer to convert changes in level to changes in the control-signal voltage with which the tachogenerator voltage is compared to obtain an error voltage for initiating operation of the induction regulator drive as the rate of outflow changes.

At any level within the proportional zone the motor is running at a particular speed, but when the outflow alters the float moves and the control signal voltage changes in proportion to the deviation of the level, the motor speed and pumping rate being adjusted to maintain the new level within the proportional zone. The pumping rate is therefore varied in accordance with the outflow and the level is maintained within closer limits than would generally be practicable with on/off control.

Another typical application of the *Asrec* system for continuously variable speed control of a type N-S motor is to maintain a constant cutting speed with a boring machine. For this requirement the table (and motor) speed has to be varied in inverse proportion to the distance of the tool from the centre of the table. In order to measure the cutting speed, the field of the motor-driven tachogenerator is supplied by a potentiometer driven by the tool-holder. As the tachogenerator voltage is then proportional to the product of field excitation and generator speed, it is therefore proportional to the product of toolholder displacement and table speed, i.e. cutting speed, with the potentiometer set to give zero output at the centre of the table.

The desired cutting speed is preset on the control potentiometer so that as the tool-holder moves away from the table centre, the tachogenerator voltage varies to give an error voltage acting to reduce progressively the motor (and table) speed to maintain a constant cutting speed.

The *Asrec* system is also applied to type N-S motor drives for textile and other beaming or reeling machines when the linear speed and tension of the material must be kept constant during the reeling operation by reducing the

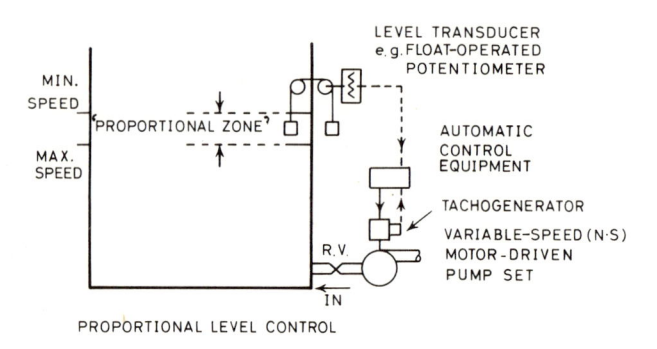

Figure 6.9. Variable-speed pump drive

rotational speed as the reel builds up. A tachogenerator driven from a measuring or idle roll provides a linear speed signal to enable the control system to reduce the motor speed continuously.

Several motors can be controlled simultaneously by the *Asrec* system when they are required to run with fixed preadjustable speed ratios. A basic speed is preset by a common control potentiometer and the necessary speed differentials are established by adjusting a rheostat in the field circuit of the tachogenerator of each motor.

Multi-motor Variable-speed Pumping Installation

The *Asrec* control system is used for the automatic control of pumping installations when with changing demand several N-S motor pump drives have to be started or stopped according to the number required, and the speed of the running units also has to be varied over a wide range. A typical arrangement is to have a 'leading' pump which, when the output required is below its capacity, runs by itself and with increasing demand is speeded up until, when it is fully loaded, a second unit is started automatically and the speeds of both units are adjusted so that they share the load. The speed of the leading unit is reduced to that set for the second unit to enable it to pick up half the total load. When both units are fully loaded a third unit is started to share the load, and so on according to the number of units installed.

Figure 6.10 is a schematic arrangement of an automatic control scheme for three type N-S motors driving pumps required to maintain a level. When

A.C. Motor Drives

only the leading pump is running, with increasing demand the float-operated potentiometer increases the speed reference voltage and the motor speeds up until the tachogenerator voltage reduces the error voltage to the value to keep the motor running at the speed necessary to meet the demand. The two voltages are compared in the discriminator 1 giving an output to regulate only the speed of the leading set. When the leading pump reaches its top speed and full capacity the induction-regulator rotor has moved to a position where a limit switch closes to initiate control action to start another pump and adjust the speeds of both pumps so that they share the load.

With the closing of the limit switch the control relay CRA operates to change the speed reference voltage by altering the ballast resistance in series with the float-operated potentiometer so that the speed-control circuit of the second pump motor is energised. The same process is repeated for the third pump and operates in reverse for shutting down pumps as the demand decreases.

A trimmer resistor is included in the tachogenerator field-winding circuit to provide for the relative adjustment of the pump speeds when running in parallel in order to match their characteristics correctly to ensure economical

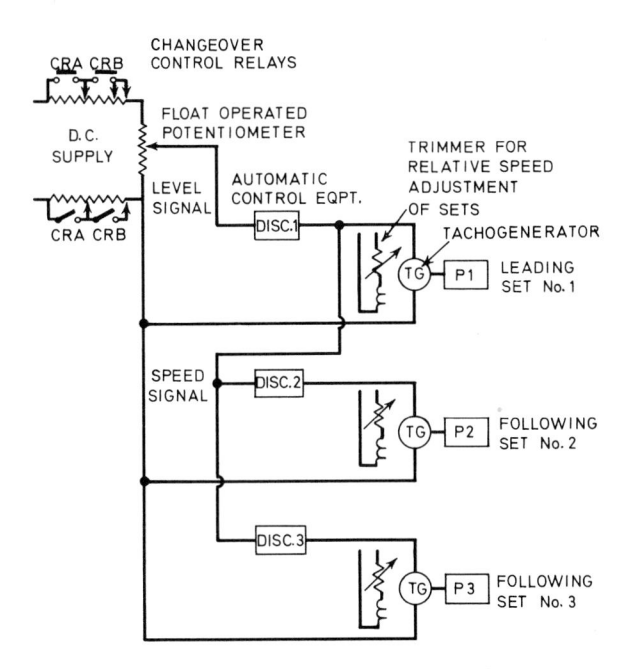

Figure 6.10. Automatic control scheme for multi-motor pumping installation

operation. For this reason the pumps may have to run at slightly different speeds to compensate for unequal wear, the condition of the piping and similar small variables.

Also developed by Laurence, Scott & Electromotors for the control of N-S motors the *Asmag* and *Ascron* systems are based on the use of magnetic amplifiers and thyristor switches, respectively. The magnetic amplifiers are self-saturating types of high-response design supplying the complete power

98   A.C. Motor Drives

to a d.c. series pilot motor with a split field to provide for precision control and rapid reversal. With the *Ascron* system controlled thyristors switch the supply to a 3-phase pilot motor (see page 25).

The *Asmag* and *Ascron* control systems have performance characteristics superior to those of the *Asrec* system, but it will be appreciated that the latter is completely adequate to meet the control requirements for many automatic drives, one of the other two systems being adopted only when it is essential for maximising the operational performance of the driven unit.

Type CH Motor Control Schemes

The speed regulation of Schrage type 3-phase commutator motors is obtained by schemes incorporating control equipment for operating the pilot motor which adjusts the brushgear rockers. Various schemes developed for the control of the type CH motors produced by English Electric-AEI Machines Ltd are based on the use of a preset speed controller.

Figure 6.11 shows schematically the arrangement of a preset speed controller which has two cam-operated switches (EF and CD) that are used in the control scheme shown in *Figure* 3.21 (page 45) in place of the 'raise speed' and 'lower speed' pushbuttons. The pilot motor driving the brushgear rockers also drives the cam assembly but this can also be rotated independently by a setting knob for selecting a required preset speed.

With the cams in the position shown in *Figure* 6.11, if the assembly is rotated slightly anti-clockwise just enough to close switch EF and the main motor is then started with the brushgear rockers in the minimum speed

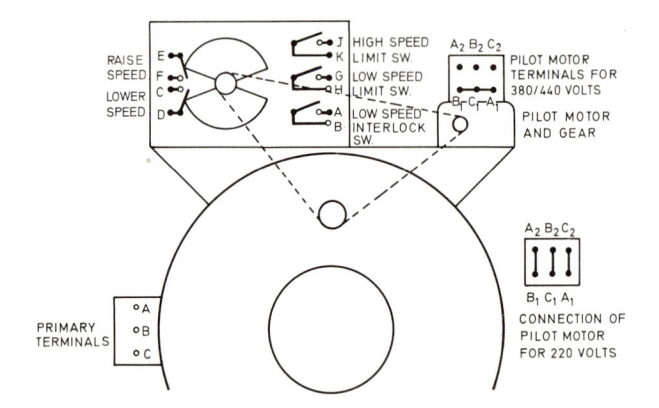

Figure 6.11. Schematic arrangement of preset speed controller

position, the pilot motor runs to drive the rockers and the cam assembly to the maximum-speed position. As the pilot motor stops when the 'raise speed' cam disengages switch EF, it follows that the operating time of the pilot motor is determined by the rotational adjustment of the cam relative to the switch which therefore determines the final rocker position and the main motor speed corresponding to that position. The appropriate adjustment of the cam assembly by the control knob presets the speed to which the main motor runs up when started, but if this speed is below the maximum it can be raised while the motor is running by operating the control knob.

Operation of the control knob in the clockwise direction while the motor is running causes the 'lower speed' cam to close switch CD so that the pilot motor runs to move the brushgear rockers to lower the speed to a value determined by the setting of the control knob and therefore the operating time of the pilot motor.

The latest design of preset speed controller produced by English Electric-AEI Machines, the type C748, is available in several versions similar in

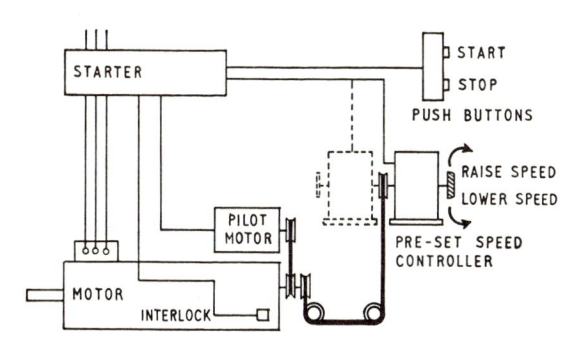

Figure 6.12. Control scheme with remote-mounted preset speed controller. The broken line shows an alternative arrangement of the controller.

principle to the controller shown in *Figure* 6.11 but with one or two cams or switches according to the application requirement. For example, the type C748–C with one cam and one switch is, in effect, an adjustable limit switch set with the motor at rest. The cam operating the switch will retain up to the maximum speed of the motor. Normal use of the controller is either to determine a base speed up to which the motor will run and which can then be over-ridden by pushbuttons, or alternatively to set a maximum speed below which pushbutton control is used. The range includes versions for mounting on the motor or remotely. *Figure* 6.12 shows a control scheme with a remote-mounted controller.

Apart from being used to preset operating speeds, the controllers are also used for the speed regulation of industrial machines, e.g. in conjunction with floating or compensating rolls to maintain constant tension in paper, cloth and textile yarns. In such cases the controller is fitted with sprockets on both shafts (*Figure* 6.13), one driven by the floating roll mechanism and the other coupled to the brushgear rocker-operating spindle. Two or more mechanically independent drives can thus be kept in synchronism by the action of the floating roll, with very light tension in the material supporting the roll as the torque required to operate the controller is small. Similarly, the speed of a pump can be controlled by the position of a float attached to the controller.

To enable a preset speed controller to be adjusted by remote control it can be fitted with a small motor operated by pushbuttons in the forward or reverse direction to raise or lower the speed of the main motor (*Figure* 6.14). The adjusting motor can also be operated under automatic control with input signals from sensing devices.

Speed regulation of the Schrage type motor can also be effected automatically by control schemes which do not include a preset speed controller. *Figure* 6.15 shows a closed-loop scheme using a tachogenerator signal voltage which is compared with a master reference speed signal to give an error signal that initiates operation of the pilot motor in the correct direction to effect the necessary speed change. When large errors occur, the brushgear rockers are driven at full speed to effect rapid correction. With small errors,

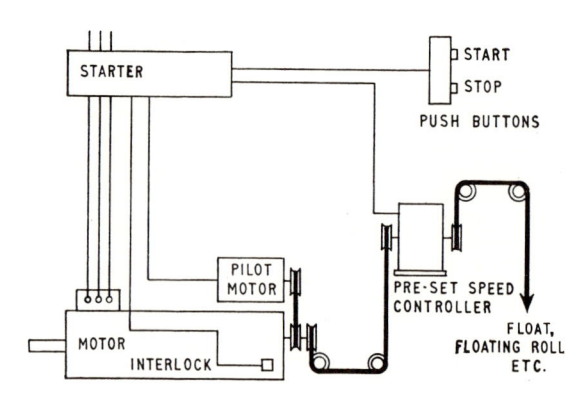

Figure 6.13. Scheme with preset speed controller adjusted automatically for industrial control functions

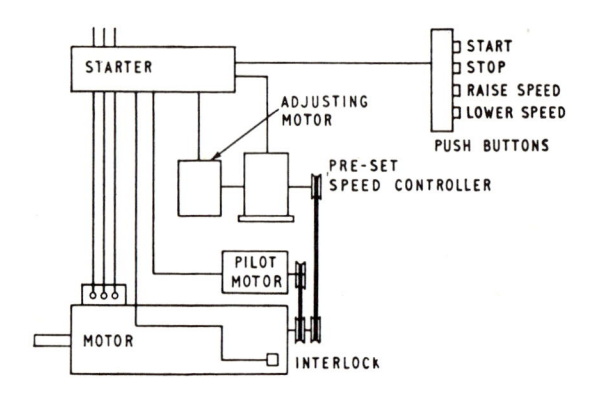

Figure 6.14. Scheme with motor-adjusted preset speed controller

the pilot motor is driven in a series of pulses so that the brushgear rockers are moved in small increments, and accurate correction is achieved without overshoot.

For a given speed setting, the inherent shunt characteristic of the Schrage motor results in a no-load to full-load speed drop of between 5–10% of the maximum speed of the motor. With a closed-loop control scheme the speed can be regulated to within $\pm\frac{1}{2}\%$ of maximum speed at any desired speed.

The system will regulate to a preset speed on a single drive, or follow a speed signal from another drive, depending on the source of the master reference speed signal. With such a scheme it is possible to maintain the speed

A.C. Motor Drives

of several motors within $2\frac{1}{2}\%$ of a master reference speed which may be derived from one of the motors or another source.

As an alternative to using a preset controller, a speed feedback signal can also be obtained from the brushgear position by means of a potentiometer also driven by the pilot motor. This method of control is cheaper than the closed-loop system because it does not require a tachogenerator, but it has

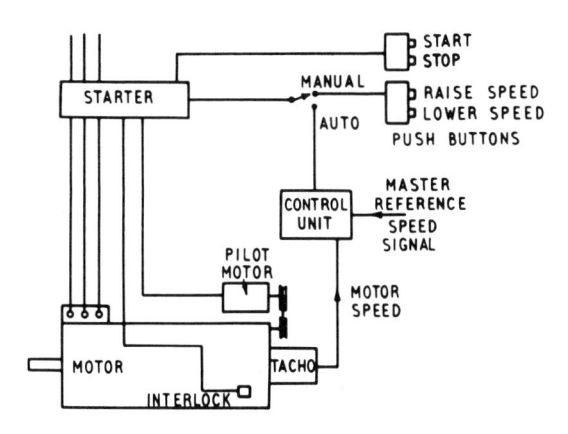

Figure 6.15. Closed-loop control scheme with tachogenerator speed reference feedback

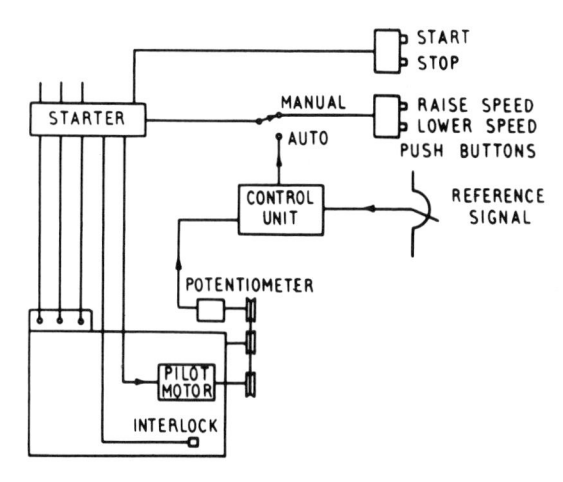

Figure 6.16. Closed-loop control scheme with potentiometer speed reference feedback

the disadvantage that the motor retains its inherent speed/load characteristic. The scheme is shown in *Figure 6.6*.

Some load compensation can be obtained with brushgear position feedback by using a current transformer connected in the motor secondary circuit. With this form of compensation, the load regulation can be made level at maximum and minimum speeds, and the worst regulation, which is at synchronous speed, is the inherent regulation of the motor.

Chapter 7

Generator-Fed D.C. Motor Drives

Certain d.c. motor drives are still supplied from a general system energised by motor-generators or rectifiers. In some cases a d.c. voltage-regulated system provides a variable-voltage supply to the armatures of a group of motors driving the units of a continuous processing line and any speed differentials necessary are obtained by regulators in each separately excited shunt-field winding circuit.

In general, however, most generator-fed d.c. motor drives are now based on a Ward-Leonard set supplying a d.c. voltage which is varied by a thyristor regulator in the generator shunt field circuit. For some drives the generator may supply two or more motors, any individual speed adjustment being obtained by motor shunt field regulators or booster generators. With the basic scheme shown in *Figure* 7.1, the generator supplies both a main motor and a reeler motor with a booster generator in series with its armature to adjust its speed relative to that of the main motor.

Ward-Leonard Drive Control Systems

Various control systems have been developed to obtain rapid and precision control of Ward-Leonard drives. A typical example, the *Velonic* system of Laurence, Scott & Electromotors, is primarily for precision speed control but it is adapted readily to control other parameters which can be represented in voltage analogue terms, e.g. torque, power and tension.

The basic *Velonic* scheme for controlling a separately excited d.c. motor is shown in *Figure* 7.2. The speed reference tachogenerator TG_1 is direct coupled to the non-drive end of the motor, and its output voltage, which is directly proportional to speed, is compared with the P_1 potentiometer slider voltage which is derived from a precision voltage-reference unit. The voltage at the junction of the summing resistors R_1 and R_2 represents the error between the desired speed, as set on the P_1 potentiometer, and the actual speed as measured by the tachogenerator.

Generator-Fed D.C. Motor Drives

The error voltage is fed into the velocity-error amplifier and the amplifier output voltage V_{ZD_1} (which is limited to a maximum value by the Zener diode ZD_1) is then fed into the current-control loop and compared by means of resistors R_3 and R_4 with a voltage analogue of motor armature current VP_2. The resultant voltage is then fed into the current-error amplifier which, unless a preset level of armature current has been reached, controls the thyristor regulator firing circuits and therefore the generator output required to maintain the motor desired speed. If R_3 and R_4 are equal, it will be seen that to the first order the level of armature current, as set on potentiometer P_2, will be controlled so that voltage VP_2 does not exceed voltage V_{ZD_1} under any operating condition. Since the current transducer

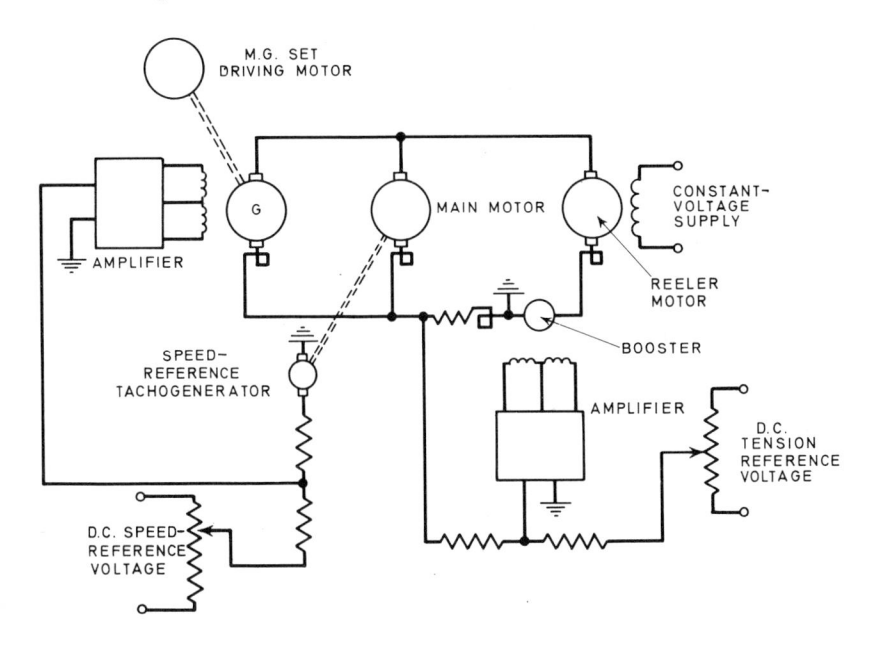

Figure 7.1. Ward-Leonard set comprising main drive motor and reeler motor with speed-regulating booster generator

gives a voltage output proportional to the armature current, this control circuit functions to limit the current to the preset value, reducing the motor torque and speed if necessary. By means of the $C_2 R_7$ stabilising circuit, a signal voltage of the correct phase and amplitude, and proportional to the rate of change of generator voltage, is fed back into the input of the current-error amplifier and this, together with the stabilising signal feedback across the velocity-error amplifier, ensures stable operation of the speed-control system over its full operating range.

When the controlled parameter is torque or current, the motor armature current can be regulated using an overriding current-control loop similar to that shown in *Figure 7.2*. The velocity summing network and associated amplifier would be replaced by a variable reference voltage proportional to the desired current level. An alternative simple form of spillover current-limit control (see page 111) can be utilised on drives not normally required to operate at their maximum current limit level, but where some form of protection is required for occasional overloads. This circuit would be arranged so that when a voltage proportional to armature current exceeds a preset bias voltage or Zener turnover voltage, a current from the armature circuit would be fed into the velocity summing network and attenuate the amplifier input signal to limit the current to the required level.

The basic circuit shown in *Figure* 7.2 is for a unidirectional drive which could be reversed by reversing the polarities of its tachogenerator and motor field. If only limited reverse running is required, the motor field and tachogenerator connections can be switched, but for rapid operational reversing it is necessary to have equally balanced double field windings on the generator fed from a push-pull amplifier.

The drive can then be reversed rapidly by switching the polarity of the reference voltage, or alternatively, where desirable, the reference voltage can

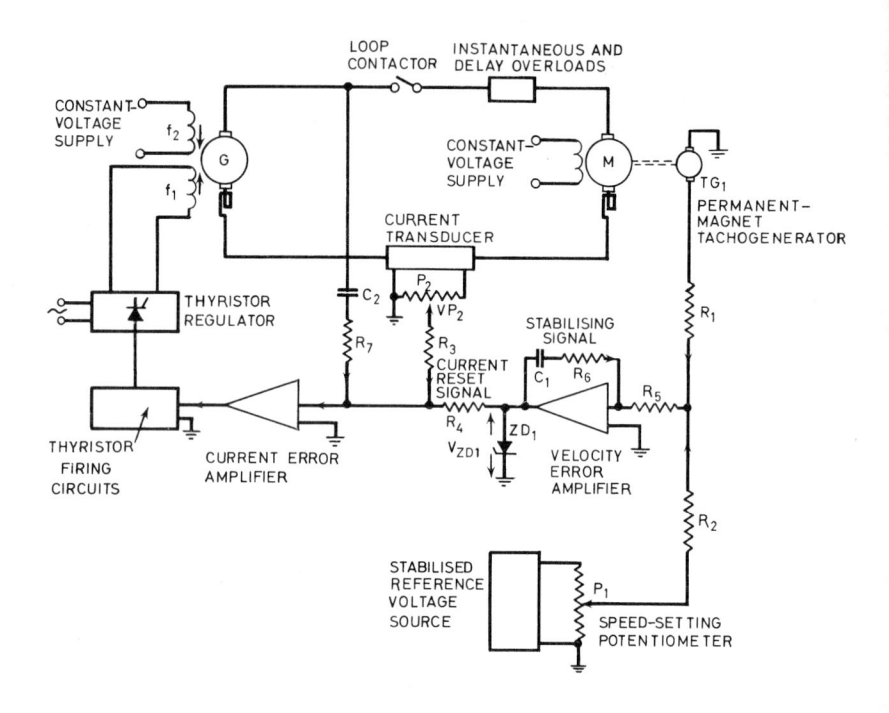

Figure 7.2. Velonic *speed-control system*

be centre-tapped so that one rotation of potentiometer P_1 would take the drive from maximum speed in one direction through zero speed to maximum speed in the reverse direction.

With overriding current-limit control *Velonic* drives can be switched directly from maximum speed in one direction to maximum speed in the reverse direction without overloading the motor or causing flashover of the commutator. The reversing or decelerating time will be governed by the output torque of the motor at the current limit level, and the load and inertia of the driven machine.

Centre-spindle Reeler Control Scheme

Typical variable-speed applications of the *Velonic* system are to centre-spindle reeler (or decoiler) drives. The scheme shown in *Figure* 5.8 provides tension feedback control with an adjustable taper tension facility.

Figure 7.3 shows a scheme for a constant-power reeler drive incorporating a rotary multiplier which functions to maintain a constant power output

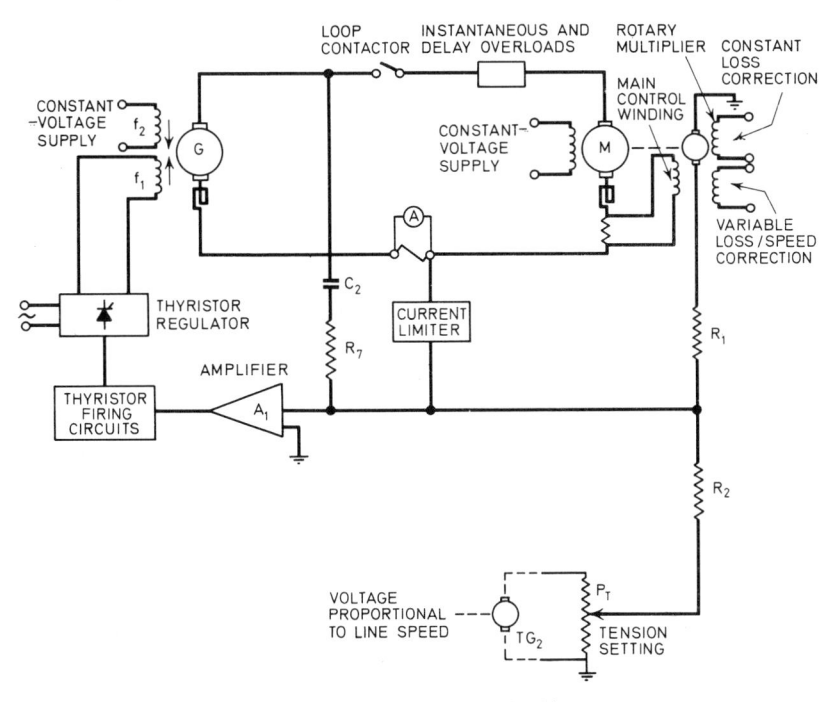

Figure 7.3. Constant-power reeler drive

from the reeler motor so that its speed falls as the coil builds up and the material linear speed and tension remain constant. The rotary multiplier is virtually a tachogenerator driven by the reeler motor and fitted with three or more separate field windings. The main control winding, connected across a resistor in the motor armature circuit, is excited with a current proportional to the armature current. The other windings are fed from loss-compensation circuits so that the resultant output voltage from the rotary multiplier is directly proportional to the power (tension × speed) transmitted to the material being wound.

The output voltage is compared, by means of a conventional summing network, with a voltage proportional to the desired tension (at a particular linear speed) as set on potentiometer P_T and the error voltage is fed into amplifier A_1 to maintain the output power of the reeler motor at the preset level. As the power transmitted to the material is proportional to its linear speed, the reference voltage connected to potentiometer P_T must also be proportional to linear speed in order to maintain the same tension over the whole linear speed range. A convenient reference voltage can be derived from a tachogenerator running at line speed and coupled to the drive of a preceding section.

Although thyristor converters are now being used for many drives previously powered by Ward-Leonard sets, the latter are still used for particular applications, and schemes engineered some years ago are of course continuing to meet the control requirements for both single-motor driven units and processing plants involving the co-ordinated control of several motors.

Planer-table Drive

The single-motor drive scheme shown in *Figure* 7.4 for the closed-loop control of a planer table was developed by AEI some years ago to utilise the then

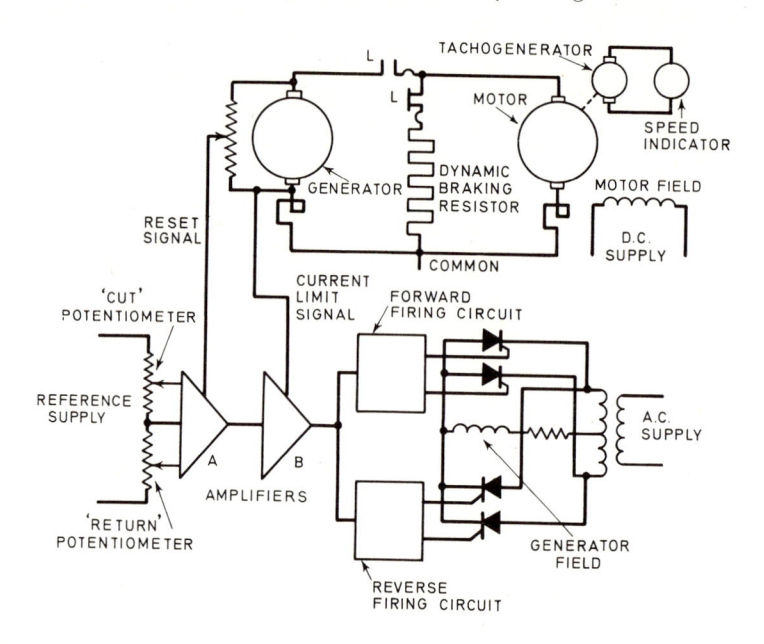

Figure 7.4. Planer table drive

new semiconductor devices as replacements for relays and contactors both to reduce the time-lags inherent in previous planer-table control equipments and to minimise maintenance.

A basic feature of the scheme is the use of a reversible thyristor amplifier for regulating the excitation of the generator field.

Reversal of the planer table is initiated by two AEI dry-reed vane switches which operate when a ferrous element attached to the table enters a slot in the switch. When the table is stopped a power-off memory device 'remembers' the direction of its motion so that it will be continued in the correct direction when the table is started again. The device comprises a transistor flip-flop circuit feeding into two magnetic cores which remain magnetised when the supply is removed.

When the start button is pressed the loop contactor L operates to connect the motor to the generator, and an input signal from either the 'cut' or 'return' potentiometer, determined by the power-off memory device, is applied to the pre-amplifier A. This signal is fed to amplifier B and then applied to either the forward or reverse thyristor firing circuits. With the thyristor amplifier operating to increase the generator voltage, the motor accelerates and a speed signal proportional to back e.m.f. is fed back into amplifier A. An overall current limit signal is used to control the loop current during acceleration. Once the motor has accelerated to the preset speed this is maintained independent of load variation because of the feedback signal.

When the vane switch at the end of the table is operated, the polarity of the input to amplifier A is reversed and the input to the generator field reduces so that the motor can be braked regeneratively to standstill at a rate determined by the current-limit feedback signal. (The dynamic braking resistor is switched across the motor only when contactor L opens.) Conduction of the second thyristor converter is held off by an inhibit circuit in amplifier B until the generator field falls to zero, thus preventing large circulating currents in the bridge circuits. A tachogenerator energises a motor speed indicator.

Crab-mounted Bucket-hoist Drive

The Ward-Leonard scheme shown schematically in *Figure* 7.5 was engineered by AEI for the operation of a crab-mounted bucket hoist for charging a pair of blast furnaces. Filled with ore and coke the bucket is hoisted vertically for 35 m until it is higher than the furnaces and the crab is then traversed horizontally along the track until the bucket is above the furnace and is lowered on to the furnace top. The bucket is of the bottom-opening self-discharging type which, on landing on a cradle on the furnace top, dumps its load on to the main bell when the lowering rope is payed out. A short time delay is allowed to ensure that the bucket is emptied completely before recommencing the series of operations to return the bucket to the charging pit.

The 186 kW (250 hp) hoist motor and the 26 kW (35 hp) cross traverse motor are supplied from a Ward-Leonard 230 kW 500 V generator. The generator field is supplied by an amplidyne exciter which is driven by a separate squirrel-cage motor also driving a constant-voltage exciter for the motor shunt fields.

A reference voltage sets the required hoist motor speed, and the motor back e.m.f. used as the speed feedback signal is obtained from a bridge-connected resistor scheme (see page 72) which compensates for the variation in armature voltage drop with current loading. The scheme ensures that the hoist speed is independent of load. Acceleration is controlled by the action of the amplidyne stability winding in series with the resistance-capacitance (RC) circuit which controls the rate of increase of the exciter output and therefore the generator voltage and motor armature current.

Any tendency for the current and, therefore, the acceleration to increase too rapidly is checked by the current flow through the *RC* circuit, which is determined by the rate of rise of the applied voltage. The dynamic braking resistors are applied automatically when the motor armature circuit contactors open (as shown in *Figure* 7.5).

Aluminium Foil Mill Drives

An interesting Ward-Leonard multi-motor drive system was engineered by Laurence, Scott & Electromotors Ltd. for two rolling mills in the factory of Aluminium Foils Ltd. The installation comprises two Loewy-Robertson four-high mills for producing aluminium foil in a width range of from 91 cm to 163 cm, from 0·6 mm stock rolling down to 0·008 mm with a tolerance of better than ±2% (±1% on final process). One mill is used normally for roughing passes and also on occasion for finishing, being referred to as the dual purpose (d.p.) mill. The second mill is used only for finishing.

Each mill has four main drives: doubling unwind, unwind, mill and rewind (*Figure* 7.6). Foil with an entry thickness above 0·035 mm is fed into the mill singly and wound up on the rewind after being rolled with a reduction of 40–70%. Thus in order to produce the thinnest foil from stock

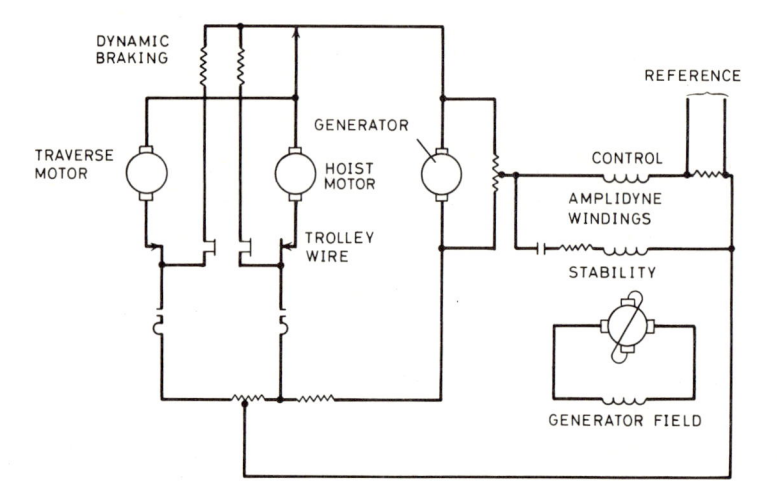

Figure 7.5. Crab-mounted bucket-hoist drive

material it may be necessary to make five passes through the mill. Below an entry thickness of 0·035 mm the foil is fed into the mill double, one strand from the unwind and another from the doubling unwind. The two thicknesses are rolled together and wound up as double foil, being later separated and rewound on another machine.

The basic requirement of the rolling process is that the rolling speed should be regulated by controlling the speed of the mill motors, and the tension of the ingoing and outgoing foil should be controlled by means of the unwind and rewind motors. One of the most important parameters of the finished product is its thickness, which must be kept to a close tolerance, and which is affected by such factors as oil conditions, temperature, tension and rolling speed.

With these mills operating up to a speed of 762 m/min (2500 ft/min) it is necessary to provide an automatic gauge control (a.g.c.) facility which initiates regulation of the mill motor speed in accordance with the principle that an increase in mill speed reduces the foil thickness. In addition, the

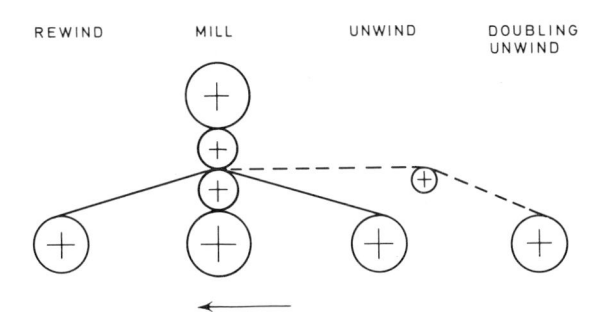

Figure 7.6. Diagrammatic representation of foil rolling mill

unwind and rewind tension must be closely controlled so that variations do not affect the gauge. For these reasons closed-loop control is essential for all the main drives.

To ensure good-quality foil the unwind and rewind stands must be as close as possible and there must not be any idling rolls between these sections. For this reason load cells cannot be used as a measure of the foil tension and therefore the unwinds and the rewinds are controlled by keeping the armature current proportional to the set tension, and the motor field flux proportional to the coil diameter, while providing accurate compensation for the motor and reel-stand losses.

All four drives on each mill have d.c. motors supplied from Ward-Leonard generators. The d.p. mill itself is driven by a separately excited shunt-wound compensated motor rated at 0/746/746 kW (0/1000/1000 hp) at 0/350/875 rev/min at 0/500/500 V, and capable of frequent working peaks of twice full-load torque. The motor speed range corresponds to foil speeds of 0/305/762 m/min (0/1000/2500 ft/min). The motor is force ventilated with filtered air from an external source.

The motor armature is supplied from an 825 kW continuous-rating compensated double-shunt-wound generator. The main shunt field is supplied from a thyristor unit and there is also a small separately excited bias shunt winding which ensures full speed control down to zero output by neutralising the generator residual field flux.

A 1044 kW (1400 hp) 3·3 kV *Trislot* induction-motor drives the generator and also the generators supplying the unwind and rewind motors.

The finishing mill motor is a force-ventilated separately excited shunt-wound compensated type continuously rated to develop 0/447/447 kW (0/600/600 hp) at 0/350/875 rev/min at 0/500/500 V. The motor is supplied from an uncompensated double-shunt-wound generator rated at 500 kW and driven by a 634 kW (850 hp) 3·3 kV *Trislot* induction motor which also drives the generators supplying the unwind and rewind motors.

Figure 7.7 shows the speed-control scheme for the mill motors. Up to the base speed of 350 rev/min the motor is rated at full torque and the motor field excitation is held at constant maximum value, speed regulation being by adjustment of applied armature voltage. Above base speed the motor is rated at constant power, and a further increase in speed up to 875 rev/min is obtained by field weakening with a corresponding loss in available torque.

A temperature-compensated tachogenerator coupled to the mill motor provides a voltage proportional to motor speed and this is compared with an adjustable reference voltage proportional to the required speed. The error

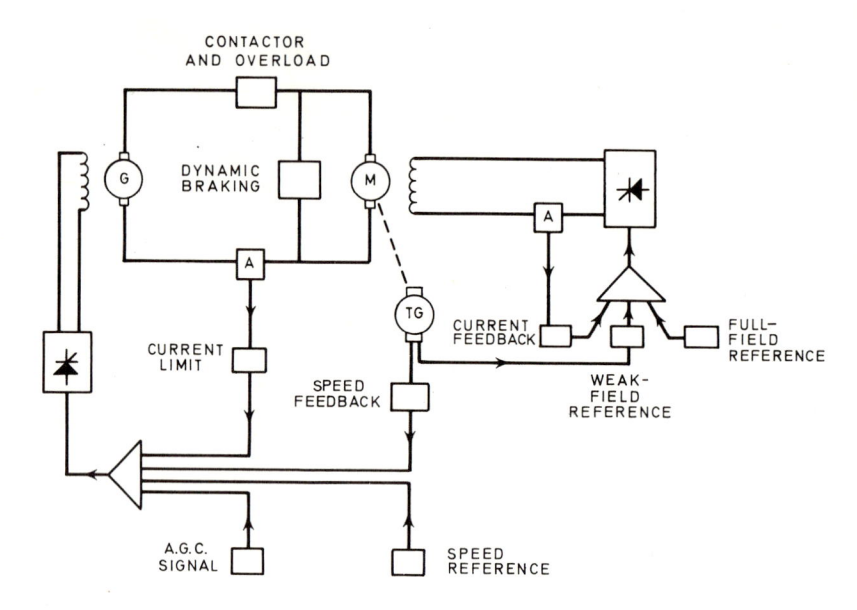

Figure 7.7. Mill-motor control scheme

voltage is fed to the input of an amplifier having a transistor input stage and a thyristor output stage, which provides a variable input to the generator field for speed regulation up to base speed.

The generator also has a small bias field supplied from a constant d.c. source and producing a flux opposing the main field and serving to cancel out residual flux so that the generator voltage can be regulated precisely to obtain inching and crawling speeds.

The motor field is supplied from a thyristor amplifier (*Figure* 7.7) which has signal inputs controlling the field current to the level required for speeds above base speed. The field current is controlled by a field-current feedback signal being compared with a reference voltage which is the difference between a fixed voltage proportional to the full-field (base-speed) current and a variable voltage obtained from a tachogenerator and used as a weak-field reference. The tachogenerator voltage is not effective below base speed because it is opposed by a bias voltage, but above base speed it acts to reduce the full-field reference and thereby to weaken the field to increase the speed of the motor.

The field amplifier control circuits form a spillover system so called because when the variable reference voltage exceeds the bias voltage it spills over to become effective as a control signal for field weakening. The bias voltage ensures that up to near base speed the full field is maintained so that the accelerating torque is the maximum permitted by the current-limit control and the run-up time is less than when field-weakening speed regulation is used throughout the whole variable-speed range. Accelerating torque above base speed is maximised by shaping the weak-field reference signal to correspond with the field-current characteristic of the motor.

Field weakening is used to increase speed above 350 rev/min, the armature voltage being kept approximately constant at maximum value, only minor adjustments are necessary to allow for changes in load, supply voltage, etc.

The operator's speed controls comprise a crawl/hold/run flick switch, a speed-setting potentiometer covering the range 762–152 m/min (2500–500 ft/min), and a crawl-speed setting potentiometer calibrated 61–9 m/min (200–30 ft/min). When 'run' is selected the motor accelerates at a controlled rate from whatever crawl speed it was running at, up to the speed preset on the main speed-setting potentiometer. The speed-control circuitry is shown in *Figure* 7.8.

Under 'run' conditions the output of the operator's speed-setting potentiometer is compared with that of a reset potentiometer driven by a d.c. pilot motor. Any difference between the positions of the two sliders produces an error voltage which is fed to a transistor amplifier, the output of which is used to operate raise-speed and lower-speed relays that switch the pilot motor to enable it to turn the reset potentiometer until the sliders are aligned and the error disappears. Coupled to the reset potentiometer is the P_3 potentiometer which provides the speed reference voltage during normal running, so that when the operator selects a new speed for the mill the change occurs smoothly at a rate determined entirely by the speed of the pilot motor.

The pilot-motor speed is determined by the armature voltage selected by a switch which therefore enables the mill to be controlled to accelerate from crawl speed to top speed in 10, 20, 30 or 60 s. The deceleration of the mill is adjustable independently and normally is set to be slightly less than the corresponding acceleration rate. When the mill control is switched to 'hold', the speed is kept constant irrespective of the position of the speed-setting potentiometer. This feature is useful for adjusting operating conditions during the initial acceleration of a reel.

The stabilised voltage reference unit providing the speed reference signal via P_3 has a smooth 200 V d.c. output which is constant to within $\pm 0.01\%$ for variations in output current and $\pm 10\%$ mains voltage. With speed-sensing and error-detecting components of an equally high standard, the overall gain of the control system is very high and the accuracy of speed control is better than 0.1% of top speed.

A current-limit unit (*Figure* 7.7) protects the motor and generator against armature currents greater than twice full-load current. The current-limit unit receives an input signal, proportional to armature current, which is amplified and passed to a biasing network which under normal conditions blocks any output voltage. When the armature current exceeds twice full-load current, the bias voltage is overcome and the output from the current-limit unit is of such a polarity that, when fed into the amplifier, it acts to reduce the generator voltage until the armature current reaches the limiting value. The current is held at this value until the cause of the overload is removed and, as the current falls to the normal value, the unit is biased off again. The system then reverts from current control to speed control.

For automatic gauge control the gauge of the outgoing foil is measured continuously by a Daystrom equipment providing a direct-voltage signal, the

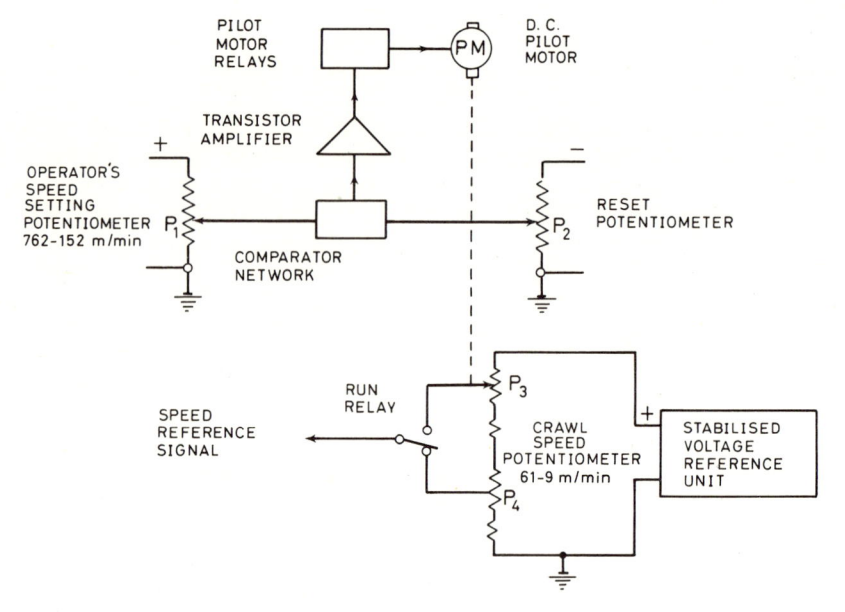

Figure 7.8. Mill speed-setting controls

magnitude and polarity of which is determined by the gauge error. This error signal is fed to the control equipment for the purpose of adjusting the mill motor speed to correct the error. Increasing the speed reduces the thickness of the foil and vice versa, although the relationship is not necessarily proportional. Accordingly the error voltage is fed via a biasing unit to the input of an integrating amplifier, the output of which provides a correction signal for the mill generator field amplifier.

The operation of the integrating amplifier is such that the output voltage increases at a preset uniform rate as long as an input voltage is present. When the input is reduced to zero the output remains constant at whatever level has been reached. The function of the biasing unit is to enable the operator to select the gauge tolerance above which the a.g.c. is required to work, two potentiometers being provided on the exit control panel for setting from ± 0.5 to 8%.

If the operator has set the required foil thickness on the control panel and set the tolerance at $+2\%$ and -2%, for example, while the foil thickness remains within these limits there is no trimming signal to the mill generator field amplifier, the motor speed being determined entirely by the operator's speed-setting potentiometer. But immediately the thickness of the outgoing foil exceeds, say, $+2\%$ error, a signal is fed to the input of the integrating amplifier, the output of which gradually increases. This output increases the output of the generator field amplifier, and the mill speed rises until the gauge of the outgoing material is again within tolerance. When this happens the biasing unit blocks the signal from the a.g.c. equipment and the output of the integrating amplifier remains constant, providing the necessary trimming signal to maintain the new speed.

If the foil thickness then decreases below -2% gauge, the opposite bias is overcome by the error signal and an opposite-polarity voltage is fed to the integrating amplifier. This causes the output, and therefore the motor speed, to reduce gradually until the foil is again within tolerance. A switch is provided in the control equipment for selecting various rates of change of the trimming signal, i.e. acceleration/deceleration of the mill under a.g.c. A speed variation of up to ± 152 m/min (± 500 ft/min) is provided by means of this arrangement.

Similar control schemes are used for the unwind, doubling unwind and rewind motors of both mills, the main difference being that the unwind and rewind on the d.p. mill have two motors per stand. On these two drives a manually operated clutch is used to couple the outboard motors when rolling heavy foil, and when this is done limit switches on the clutch select the appropriate contactor for connecting the two motor armatures and shunt fields in series (*Figure* 7.9). The outboard motor is unclutched for maximum accuracy of tension control at low tensions.

On the d.p. mill the unwind and rewind drives each comprise two separately excited shunt-wound mill-type motors, mounted on a common bedplate and coupled by means of the clutch. Each motor is continuously rated to develop 93/93 kW (125/125 hp) at 215/715 rev/min at 190 V; corresponding to coil diameters of 136 cm (53·5 in) and 41 cm (16 in) respectively when running at a foil speed of 914 m/min (3000 ft/min) on the rewind. The unwind motors drive through a 2 : 1 stepdown gearbox to give a top foil speed on the entry side of 457 m/min (1500 ft/min). In both cases speeds down to zero are obtainable against constant torque loading by armature-voltage control.

Each tandem pair of motors is supplied in series from a single 215 kW double-shunt-wound Ward-Leonard generator.

In the case of the finishing mill, the unwind and rewind are each driven by a single motor developing 93/93 kW (125/125 hp) at 215/715 rev/min, identical to one of the pair of motors on the d.p. mill. The generators are each rated at 107 kW.

Figure 7.9 shows the control scheme for the d.p. mill unwind and rewind tandem motor drives. The scheme is based on the principle (see page 75) that, since the foil speed is determined by the speed-controlled mill motor, it follows that if the rewind-motor e.m.f. is kept proportional to foil speed the motor field flux will always be proportional to the coil radius. Hence, if the flux is proportional to the coil radius, the foil tension will be proportional to armature current.

Armature-current control is achieved by taking a measure of the actual armature current and comparing this with a signal proportional to the required tension, which is taken from the stabilised voltage reference unit via an operator's tension-setting potentiometer. The error signal is fed to the input of a thyristor amplifier which supplies the generator main field and thus adjusts the output voltage to keep the current at the required value. If the mill is stopped with the unwind and rewind on tension control, the set tension is maintained with a maximum limit of about 70% of full load. A small bias field on the generator is fed from a fixed d.c. source to provide a flux to neutralise residual flux—as with the mill motor generator.

In addition to controlling the armature current to maintain the required tension, it is also necessary to allow for the losses of the motor and the reel

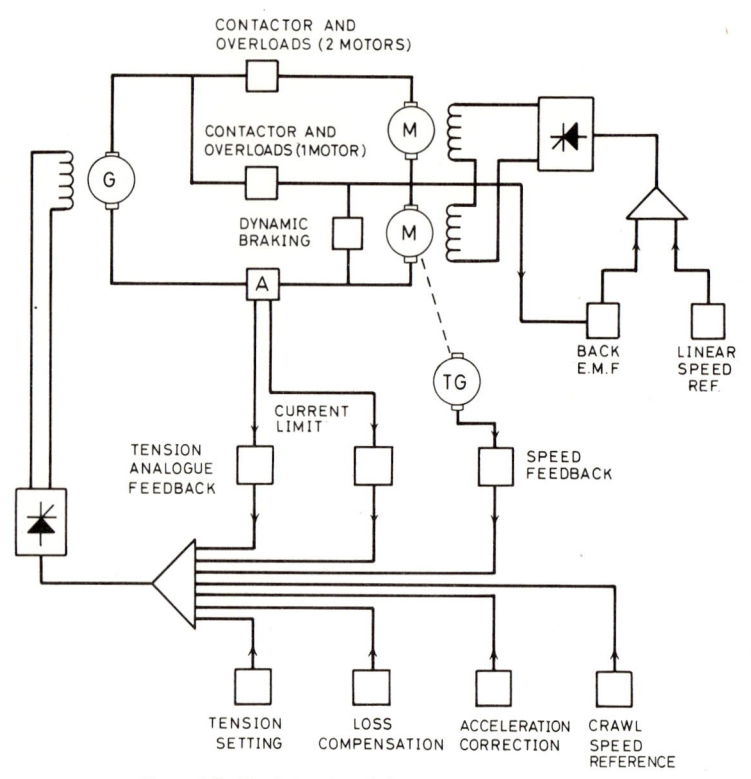

Figure 7.9. Unwind and rewind motor control scheme

stand so that the armature current provides sufficient torque to overcome the losses apart from the torque providing the foil tension. The losses are allowed for by feeding a signal into the input of the generator amplifier which exactly balances the feedback signal from the armature circuit when zero tension is set, i.e. it exactly balances the loss current signal.

The losses which have to be allowed for include the motor iron losses, which vary with field strength and speed, and the windage and friction of the motor and reel stand, which also vary with speed. Consequently, in order to obtain the correct compensation for all conditions, four separate signals are combined and fed into the amplifier input. It should be noted that the copper losses do not have to be compensated since it is the motor e.m.f. and not the applied voltage which is kept proportional to foil speed. The current control scheme includes a specially designed current transducer which enables the required accuracy of ±0·5% to be obtained.

Before the mill and rewind are threaded up at the beginning of a pass, the unwind and rewind cannot be run under tension control so a facility is provided to enable these two drives to be run under speed control for threading, the foil speed at the unwind and the rewind being controlled to be the same as that at the mill. In addition, a trimming control enables the operator to trim the speeds individually and a small amount of regulation is provided so that, when the foil is pulled tight, the rewind speed reduces automatically and the unwind speed increases to prevent the foil breaking under excess tension.

The object of the field control is to make the motor flux proportional to coil radius by keeping the motor e.m.f. proportional to foil speed, but as there are no rollers between the mill bite and either the unwind or the rewind, there is no facility for driving a tachogenerator to obtain a linear speed reference. Also, the mill tachogenerator cannot be used as a direct measurement of linear speed at the reel stands since the ratio of mill speed to unwind or rewind speed varies over the speed range.

For example, if rolling is being carried out with a 50% draft it follows that the unwind speed is always exactly half that of the rewind speed, but the mill speed lies between the two and varies with several parameters. For instance, when the mill is rolling at 305 m/min (1000 ft/min), the unwind speed might be 168 m/min (550 ft/min) and the rewind speed 335 m/min (1100 ft/min), but when the mill speed is doubled to 610 m/min (2000 ft/min) the unwind speed might be 366 m/min (1200 ft/min) and the rewind speed 731 m/min (2400 ft/min). To complicate matters, with a different thickness of foil the forward slip (ratio of rewind speed to mill speed) would be different again. In designing the scheme it was therefore decided that the linear speed at the reel stands would be computed by measuring the coil radius and multiplying it by the angular velocity.

In the case of the rewind coil it was feasible to make use of a creasing roller which presses against the outside of the coil and is carried by a pair of arms pivoted on the mill housing. A potentiometer geared to a pinion coupled to the creasing arm pivot, and with a slider calibrated in terms of coil radius provides a master reference for a servo, which in turn provides a measure of coil radius for several different circuits. The circuitry for the coil-radius servo is shown in *Figure* 7.10.

Occasionally the creasing roller is not used, e.g. on first passes with thick foil, but an alternative form of control available for the coil radius servo uses the output of the mill tachogenerator, multiplied by a preset average value of forward slip, as a linear speed reference.

To provide a control for the unwind, since there was no inherent means of obtaining a measure of coil radius, it was necessary to add a device designed expressly for this purpose. One restriction placed on the design of such a device was that no mechanical roller, wheel, etc., should touch the foil as this would spoil the surface. To meet the control requirement the pneumatic measuring device shown in *Figure* 7.11 was designed. The compressed-air supply is fed to a light vertical probe, having a nozzle at the lower end, which is positioned above the top of the unwind coil of foil but is prevented from touching the coil surface by the cushion of escaping air.

116   *Generator-Fed D.C. Motor Drives*

The probe is free to move vertically in guides, a rack on its upper section being meshed with a pinion coupled to a helical potentiometer, so that when the coil diameter changes, the probe moves vertically in sympathy and the potentiometer is turned to produce a measurement of coil diameter. A switch on the entry control panel enables the operator to retract the probe and swing the supporting arm, pivoted at A, out of the way for threading the mill.

Like the rewind, the unwind coil radius servo has a standby control network which entails using a speed reference signal, obtained from the rewind linear speed via an operator's draft setting potentiometer. The doubling unwind servo uses a speed reference obtained from the main unwind equipment. In addition, all the unwind drives are provided with coil-diameter setting potentiometers so that the servos can be set correctly to provide the appropriate field strength for threading the complete range of incoming coils.

Having obtained a measure of foil speed at the unwinds and rewind, this is used as a reference for the field control, as shown in *Figure* 7.9. A signal proportional to the motor back e.m.f. is compared with this reference and the error is fed to a thyristor amplifier, the output of which is used to supply the

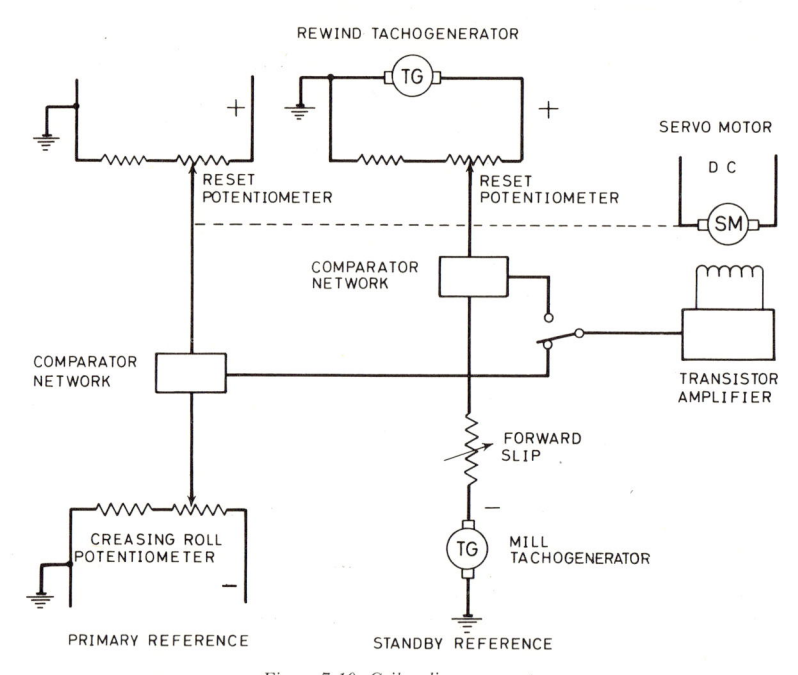

Figure 7.10. Coil-radius servo system

motor field. With this arrangement the motor field flux is maintained proportional to the coil radius as the latter changes during the course of a pass.

A signal proportional to motor back e.m.f. (E) is obtained by summing together a signal proportional to the motor terminal voltage (V) and another proportional to the volt drop in the motor (IR): i.e. $E = V \pm IR$.

The *IR* drop of the motor is composed of the armature *IR* drop, the interpole *IR* drop, the brush drop and the *IR* drop of a ballast resistor used for the measurement of armature current. Since direct measurement of the volt drop in the rotating armature is not possible, it is usual to measure the drop in the interpoles and scale up the signal accordingly. A disadvantage of doing this, however, is that when the motor current is increased the relatively small mass of copper in the interpoles warms up faster than the much larger armature winding so that the ratio of their resistances alters with changes in the operating current.

In this instance the problem was overcome by taking the *IR* drop signal across both the interpoles and the fixed ballast (which is not affected appreciably by temperature) and choosing the resistance of the latter such that

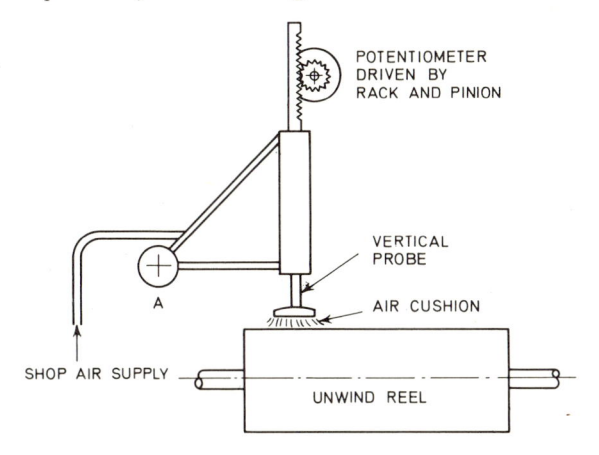

Figure 7.11. Pneumatic coil-radius measuring device

the change of signal volt drop with machine temperature is very close to the change in overall volt drop of the motor, the tolerance of the e.m.f. signal due to this effect being $\pm 0.2\%$.

One disadvantage of the above field control scheme is that, at low motor voltages, e.g. at inching and crawl speeds, random changes of ± 1 V in the brush drop would produce considerable errors in the back e.m.f. signal and hence the field strength. For this reason an alternative field control system was included for the unwinds and rewinds at low speeds, the changeover taking place at the closest steady running speed to 152 m/min (500 ft/min). This alternative system controls the motor field current to the required level dependent on the coil radius. A potentiometer coupled to the coil radius servo is graded and calibrated in terms of motor field current, and the output from this is compared with a measure of actual field current, the error signal being fed to the input of the field thyristor amplifier.

Although adequate for control below normal running speed, the field-current control system is not suitable for use at higher speeds because of errors due to variations in flux brought about by armature reaction and, at weak fields, hysteresis.

When the unwind and rewind drives are running at a constant speed the motors are controlled to produce sufficient torque to provide the necessary tension and overcome the losses. However, when the mill is accelerating or decelerating additional torque is required to accelerate or decelerate the inertia of the foil coils, motors and reel stands so that unless a signal into the control system initiates a change in torque of the appropriate amount, the required acceleration or deceleration will not be achieved.

The accelerating torque required is proportional to the rate of acceleration and the total inertia, the latter comprising two parts: the inertia of the motor and reel stand, which is constant, and the inertia of the coil, which varies with both coil diameter and foil width. Thus in order to provide exactly the right accelerating torque for all conditions of acceleration, coil diameter and foil width, an additional signal is fed to the input of the unwind and rewind generator field amplifiers, as shown in *Figure* 7.9. The value of signal current is given by

$$i_a = a\left[K_1 WR^3 + \frac{K_2}{R}\right]$$

where i_a = correction current
a = angular acceleration of reel
W = face width of coil
R = coil radius
K_1, K_2 = constants

In order to obtain this correcting signal the circuit shown schematically in *Figure* 7.12 is used. The output voltage of the motor tachogenerator, which

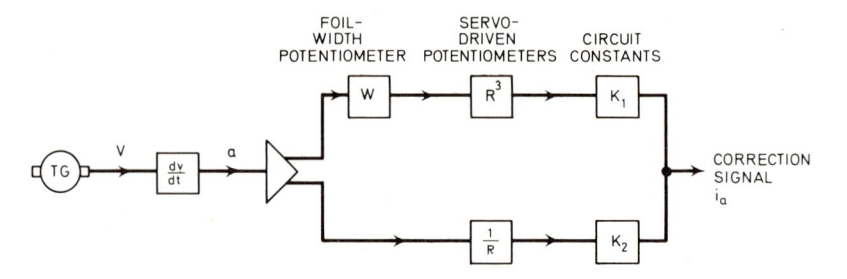

Figure 7.12. *Rewind and unwind inertia compensation circuit*

is proportional to the rotational speed of the reeler, is differentiated using an RC network, and the resultant voltage proportional to the acceleration (or deceleration) a is fed to a transistor amplifier. The amplifier output is fed to two parallel circuits, which produce signals for the required variable and fixed values of the accelerating torque.

To obtain the variable signal a voltage proportional to the acceleration is supplied to an operator's potentiometer calibrated in terms of foil width, and then to a pair of servo-driven potentiometers, which are graded and connected to produce an output proportional to R^3. The other part of the correction signal is obtained by passing the amplifier output to another servo-driven potentiometer arranged to give an output proportional to the reciprocal of coil radius. When the two signals are summed together the result produces the required current to change the motor torque by the appropriate amount.

The main drive protective devices include overloads, overvoltage and overspeed relays, thermostats for motors and generators, generator-field overloads, motor-field failure relays, anti-reverse protection for the unwinds, web-break detectors and clutch-failure protection for the d.p. unwind and rewind drives. The scheme also includes inching controls for both directions, dynamic braking for use in the event of an electrical fault, and four different degrees of regenerative braking for stopping normally, after a web break, in an emergency and in the event of a fire on the mill.

The dynamic braking is required because with an electrical fault the motors are disconnected automatically and there is no closed-loop circuit for regenerative feedback.

Normally the mill motor is braked regeneratively during run down and deceleration is controlled in conjunction with that of the unwind and rewind drives to maintain the tension within acceptable limits.

Under normal running conditions, the rewind drive motors, but it may regenerate when decelerating under light tension.

The unwind drive normally has to maintain tension in the strip so that it is regenerating, i.e. functioning as a drag generator. It is likely to motor only when the strip is being accelerated by the mill and to achieve a light tension a motoring current is necessary to overcome frictional losses and to accelerate the total inertia of the drive.

It will be noted that, since the control system uses a signal proportional to motor back e.m.f., this is proportional to the terminal voltage minus the armature circuit volt drop when a drive is motoring; and proportional to the terminal voltage plus the armature circuit volt drop when the drive is generating (see the formula on page 116). In effect, the back e.m.f. of the unwind machine is the voltage it generates normally when functioning as a drag generator.

Chapter 8

Static-Converter-Fed D.C. Motor Drives

The control schemes for drives supplied by mercury-arc converters or thyristor converters are similar in principle since in both cases the output is regulated by variation of the firing angle and a rapid reversing requirement is met by basically similar systems. In general, however, as thyristor-converter drives are a later development the control schemes utilise more advanced techniques. Also, as the thyristor was originally only a low-power device the earliest applications were restricted to small single-motor variable-speed drives. Subsequent developments have enabled thyristor converters to be used for individual drives requiring several thousand kilowatts.

The most recently developed mercury-arc converter drives are controlled by schemes utilising transistors and thyristors to maximise the rate of response to control signals. Such schemes are used with single-converter reversing drives using either an armature connection reversing switch or reversal of the motor field to obtain a reversal of the motor torque.

Mercury-arc Converter Reversing Drive

Figure 8.1 shows a basic scheme developed by English Electric using an air-operated switch for reversal of the armature connections. By comparing the voltages proportional to desired and measured speed the outer speed-control loop detects by a change in error voltage that a reversal of torque is required. The forward-speed amplifier is then caused to give a negative output which phases back the converter grids and stops the flow of armature current. The negative-speed error, plus a signal indicating zero armature current, causes the air-operated switch to change over the armature connections. An inhibit signal then prevents an output from the forward-speed amplifier and the reverse-speed amplifier is allowed to provide an output of reverse polarity into the current control circuit so that control of the motor torque is regained for reversal and acceleration to the desired speed. Deceleration and acceleration are controlled by ramp generators shown by the symbol in the blocks representing the forward- and reverse-speed amplifiers.

With the control scheme dead times of 250 ms are obtained on drives up to 7460 kW (10 000 hp) using an air-operated switch energised through a thyristor amplifier to amplify the small actuating signals with minimum delay. The switch is fitted with interlocks to prevent short-circuiting of the motor in the event of a mechanical failure.

Aluminium Strip Finishing Mill Drive

Figure 8.2 shows a mercury-arc converter scheme engineered by AEI for the main stand reversing drive of an aluminium strip finishing mill. Speed

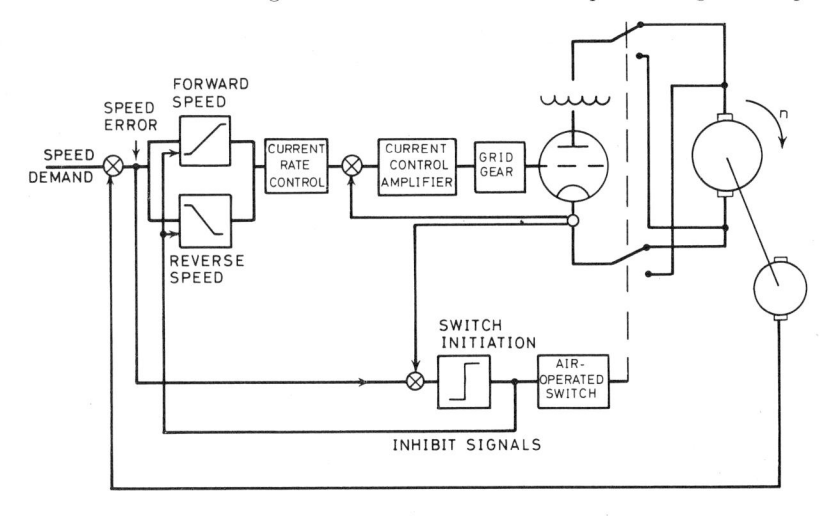

Figure 8.1. Mercury-arc converter scheme for reversing drives with air-operated switch for reversing armature connections

is varied by an input to the grid control gear from a phase-shift amplifier receiving the output from a 1000 Hz signal mixing magnastat (see page 82) with control windings energised by signals from the control devices shown in *Figure* 8.2. Co-ordinated control of the main and other drives is ensured by the speed-setting control providing a voltage signal to a *Logicon* static switching system controlling the sequencing of the mill operations.

The speed reference for the phase-shift amplifier is obtained from a static master reference unit which gives a controlled rate of change of output whenever the input is varied. The operating speed of the mill is determined by the input to the ramp generator which is preset by a potentiometer. Operation of a 'run' pushbutton produces an output signal in ramped form to the magnastat speed reference winding causing the converter control grids to be phased forward to give the voltage output required to obtain the selected speed. The speed is then regulated automatically in accordance with the signal inputs to the magnastat control windings.

Speed is raised up to base speed by increasing the applied armature voltage, and above base speed by motor shunt field weakening. The field is supplied

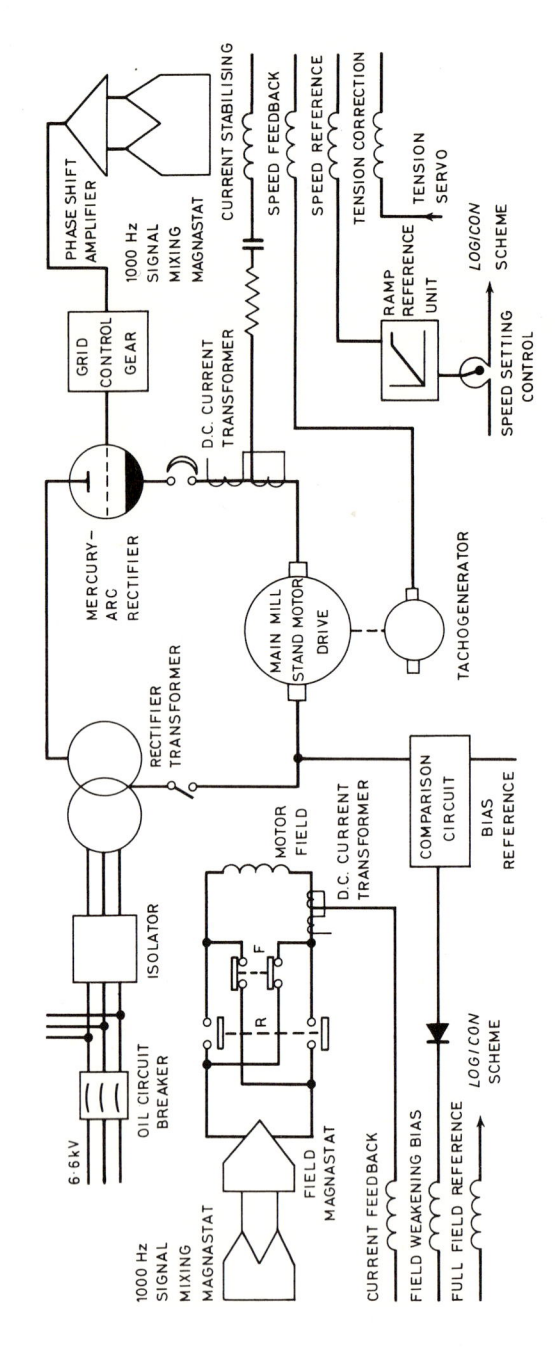

Figure 8.2. Mercury-arc converter reversing drive with field reversal scheme

from a magnastat, the output of which is controlled by the input from a signal mixing magnastat. Automatic control of the field excitation over the full speed range is obtained by the spillover method. The applied armature voltage is compared with a fixed reference voltage representing the base-speed armature voltage but below base speed there is no signal into the magnastat field-weakening bias winding. As the armature voltage is increased to accelerate the motor, when it reaches a value slightly above the base-speed reference voltage, current flows in the winding to reduce the output of the magnastat to the field winding, so increasing the speed of the motor.

The applied armature voltage is then held at the maximum value while the field strength is reduced as the motor accelerates to the preset speed. Up to this speed the armature current is maintained at a value providing for the accelerating torque required while field weakening continues. At the preset speed the field current is stabilised but it varies automatically with an adjustment of speed in the field weakening range.

With spillover speed control the mill operator is not required to adjust the field excitation so that the motor cannot be started from rest with a weak field or operated at low armature voltage with a high proportion of field weakening. Consequently, for a given speed optimum torque is obtained.

During acceleration a signal from a tension servo maintains the speed differential between the main motor and the hot coiler motor which is necessary to ensure correct strip tension. On deceleration a contactor connects a loading resistor (not shown in *Figure* 8.2) across the armature to brake the motor dynamically. Operation of the contactor is initiated by a deceleration signal from the static switching system.

Armature rotation is reversed by contactors changing over the polarity of the motor field. The changeover from forward to reverse field is sequenced in the static switching system to ensure that before the signal to the full-field reference winding of the magnastat can be removed the motor must be at rest with the rectifier grids fully retarded to give zero armature voltage. The reference signal is re-applied after the contactors have reversed the field. Only full field is available in the reverse direction.

Thyristor-converter Drive Units

Although thyristor-converter drive equipments have to be specially engineered for some applications, in many cases the requirements are met by using for each drive a standard self-contained d.c. supply and control unit ready for connection to the a.c. mains, the d.c. motor and the devices providing the incoming control signals. A typical unit is the Thorn Automation *Stardrive* available in a range for 3-phase supplies from 7·46 kW (10 hp) to 1864 kW (2500 hp) and upwards in non-regenerative and regenerative versions.

The basic components and circuits of a non-regenerative *Stardrive* are shown in *Figure* 8.3. Speed regulation is by armature-voltage variation with constant shunt field current supplied from a rectifier. The normal speed range is 20:1 below base speed but wider speed ranges can be provided. A tachogenerator feedback is shown in *Figure* 8.2, but an armature-voltage feedback version is available with armature circuit *IR* drop compensation at all speeds and compensation against speed variations due to change of field current. The speed holding accuracy of armature-voltage feedback is below that of tachogenerator feedback which can ensure a load/speed regulation of better than 1% with load changes from zero to full load independent of supply-voltage and temperature changes. The type of feedback used depends on the speed-holding accuracy required. All the *Stardrive* circuits are compensated for variations of motor load, supply voltage and ambient temperature.

An adjustable current-limit feature included as standard uses an a.c. current-transformer signal into the control amplifier. According to requirements, the *Stardrive* can be supplied with various combinations of manual controls for starting, stopping, inching and reversing, and with dynamic braking for forward only or reversing drives, and inching.

The *Stardrive-R* regenerative unit is for reversing drives where speed and/or torque have to be controlled with the minimum dead time and braking is

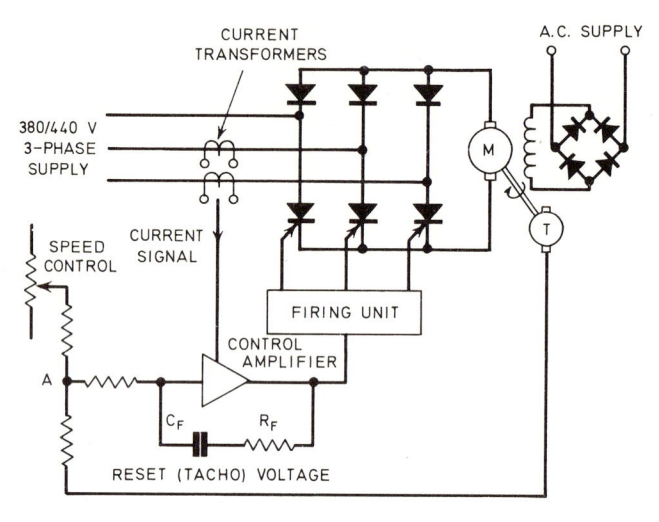

Figure 8.3. Stardrive *non-regenerative unit*

applied by the motor generating into the supply when it is being driven by the load. The variable-voltage armature supply is provided by two thyristor bridges connected in inverse parallel and individually controlled by firing units. Changeover from one bridge to the other is initiated at zero current by a control logic system in response to the input signal calling for reversal of rotation.

Tyre Tread-section Production Line Drives

Stardrives are used for both single-motor and multi-motor installations. *Figure* 8.4 is a simplified diagrammatic layout of a tyre factory line operated by nine *Stardrive* units and producing the tyre tread section. The line includes an extruder, a sidewall calender, a cushion back calender and extruder, a

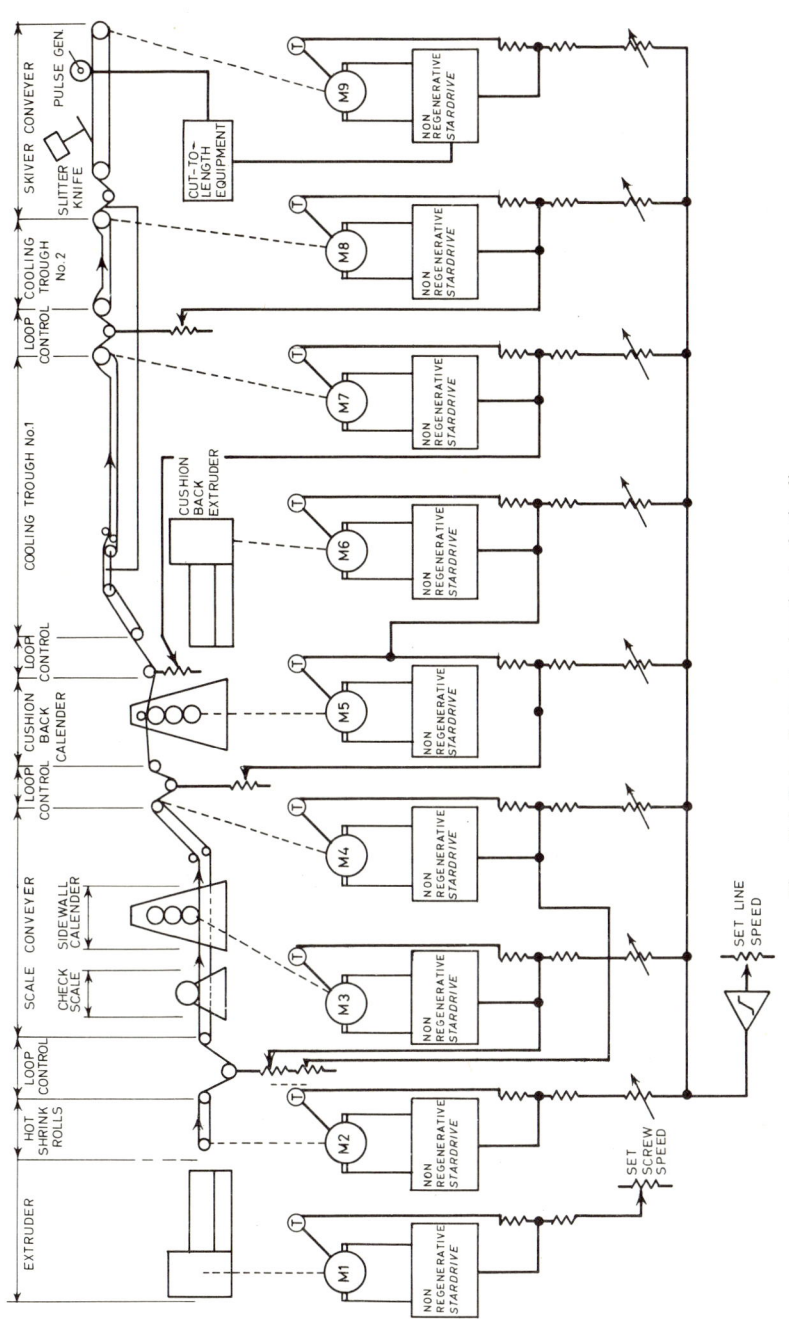

Figure 8.4. Drives for tyre tread-section production line

cutting equipment and associated conveyers. The basic tyre tread section is formed by the first extruder from strip rubber. To obtain the correct extruded section shape it is necessary to make fine adjustments of the extruder screw speed which, once set, must be held within close limits. This facility is obtained by using a non-regenerative *Stardrive* with a speed range of up to 10:1 to allow for various die sizes.

After leaving the extruder die plate, the hot and plastic tread section is supported by conveyers, two of which pass through cooling troughs. The conveyers transport the continuous length of tread section through the process line at speeds allowing it to cool and shrink during the time taken to reach the skiver conveyer where it is cut into lengths ready for assembly with the other components to form the complete tyre. The speeds of the conveyers must be adjusted and maintained relative to one another to allow for the change in length as the tread section shrinks as it travels along the line.

Although not shown in *Figure* 8.4, a typical tread line also includes equipment for applying a thin film of cement to the underside of the section ready to receive a thin layer of cushion stock prepared on the cushion back extruder and calender.

The basic operating speeds of the *Stardrives* M2–9 must be co-ordinated with the extruder screw speed, which is set by the operator who also sets the speed of the M2 drive equal to or slightly higher than the extrusion rate as necessary to produce a suitable tread section from the extruder die. The speed set for the M2 drive is also the line speed reference for the other drives. The line speed signal to each drive is adjusted by a rheostat in each speed-control circuit, and the basic speed is held by the reset feedback signal from the tachogenerator, but in addition a signal from the dancer-operated potentiometers trims the speeds to keep the loops constant. In effect, the trimming signal voltage increases or decreases the line speed reference voltage in relation to the tachogenerator voltage so that the drive speed rises or falls until the error voltage attains its steady-state value.

The tread section is cut into lengths on the skiver conveyer. While the conveyer is moving, each length is measured by a measuring wheel operating a digital pulse generator giving an input to the cut-to-length equipment. When the equipment has received the number of pulses representing the required length, the conveyer is stopped and the tread section is cut by the slitter knife. To keep the line operating while the skiver conveyer is stopped a loop is maintained between it and the preceding conveyer, but to prevent the loop growing, the skiver conveyer is accelerated rapidly above line speed after each cut.

A regenerative *Stardrive* is used for the skiver conveyer to provide for rapid braking to standstill, and to meet the control requirements of the duty cycle it has to vary the motor speed over a wide range rapidly and precisely. For any given line speed and tread length there is a fixed period of time between the completion of one measuring and cutting cycle and the next. With maximum line speed and tread length the cycle time may be only 4 s with a knife traverse time of about 1 s. The drive unit must therefore accelerate the conveyer to a speed allowing the measured length of tread to be advanced past the cutting knife, and then stop the conveyer in the minimum time, consequently the times available for acceleration and deceleration are relatively short.

Static-Converter-Fed D.C. Motor Drives

When a slow-down signal is received, the conveyer is first braked down from maximum speed to a preset crawl speed and then a stop signal initiates braking to a final crawl speed and standstill. As the maximum speed of the conveyer is a function of the tread length required, so that the point from which deceleration is initiated varies, the control scheme is designed to give the slow-down signal earlier as the maximum conveyer speed is increased. In addition, to compensate for any slight inaccuracies in the system, a definite time is allowed for running at crawl speed so that the final stop is always made from a fixed speed reference.

Another application of *Stardrives* in a tyre factory is shown in *Figure* 8.5, which is a diagrammatic layout of the plant producing the beads that hold

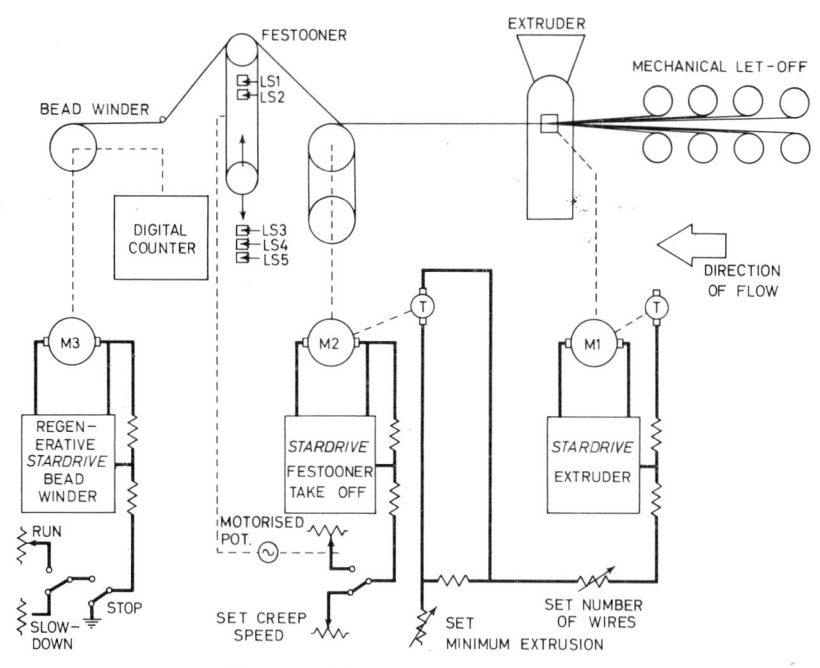

Figure 8.5. Drives for tyre bead production plant

the tyre in the rims of the wheels. The beads are formed from strands of wire which are coated with rubber and then wound in a multi-strand loop.

From the mechanical let-off the wires are hauled through the extruder by a drum driven by the M2 motor. With the control scheme shown the extruder motor runs at a speed that is based on a preset minimum speed and modified by the signal from the M2 tachogenerator circuit and the rheostat which is set according to the number of wires being used. The M1 and M2 motor speeds are maintained at a preset ratio by the signal from the M2 tachogenerator. The drum driven by motor M2 feeds the wire into the festooner at a rate determined by the bead winding operation. The M2 motor speed reference is adjusted automatically by a motorised potentiometer, which is used by the operator to preset a desired speed before starting a production run.

Wire is fed into the festoon at line speed but it is pulled out by the bead winder at about twice the line speed for the purpose of forming the bead loop. The bead-forming duty cycle necessarily includes a period when the winder is stopped so that with a constant line speed enough wire must be stored in the festoon to provide the length required when the winder re-starts.

While the correct line speed is maintained the height of the festoon oscillates about a mean position. If the line speed falls the festoon carriage (carrying the lower roll and keeping the wire taut) moves upwards until it engages the limit switch LS2 which initiates operation of the motorised potentiometer to increase the speed of motor M2. At the same time a second contact on the limit switch initiates an adjustment of the motor speed reference to obtain a slightly higher constant line speed, consequently the carriage disengages the limit switch as the festoon length increases and returns to its normal position. With this control action a decrease in line speed is compensated by a positive increase in the M2 motor speed, but if it is not sufficient to restore the carriage to its normal position, the control action is repeated. If the carriage rises until it engages the ultimate limit switch LS1 the bead winder is stopped.

The lower limit switches operate if the festoon carriage moves downwards from its normal position. Closing of the LS3 switch reduces the speed of the M2 motor, and if the festoon continues to grow the LS4 switch causes the motor to run at crawl speed. The ultimate limit switch LS5 stops the motor.

A regenerative *Stardrive* is used to meet the control requirements of the bead winder duty cycle which starts after the coated bead wire is clamped on to the bead former. The drive then accelerates to a preset fast speed and runs for the number of revolutions required to wind the bead. After a preset count of revolutions a slow-down signal causes the drive to be braked down to a preset low speed and a stop signal at the required number of revolutions stops the winder. The incoming bead wire is sheared off, the former is collapsed and the bead is removed.

The standard thyristor converter units available from several manufacturers are used for a variety of drives but in some cases complete power-supply and control schemes have to be developed in order to optimise the performance of the plant.

Steel-strip Processing Line Drives

To meet the special control requirements of the steel strip processing line shown in *Figure* 8.6, a few years ago GEC (Engineering) Ltd. developed the thyristor converter drives and control schemes shown in *Figures* 8.7 to 8.10. The double-converter reversing drives utilise the so-called suppressed figure-8 system of connection shown in *Figure* 4.10 (page 62). The line is used to bright-anneal and pickle cold-reduced stainless-steel strip, from 0·56 to 1·26 m (22 to 50 in) wide and 0·25–3·18 mm (0·01–0·125 in) thick, at a maximum speed of 24·4 m/min (80 ft/min).

The control requirements were met by using drive units in three categories: speed-controlled and with good regulation characteristics for controlling strip speeds; speed-controlled with a tension control for controlling section tension; and tension-controlled for uncoiling and coiling the strip.

As the line operates continuously, a special feature is the provision of accumulators to hold a sufficient reserve of strip to maintain processing while the end portion of the coil being processed is held for the welding-on of the end of a new coil, and to store surplus strip while a completed coil is being removed.

In the entry section (*Figure* 8.6) the coil is loaded on to one of the two mandrel-type decoilers, fed through the associated pinch roll and flattener unit and the leading end is sheared ready for welding to the end of the previous coil. The primary purpose of the decoiler is to keep a constant back tension on the strip fed into the process. During shearing of the tail end and subsequent welding of the coils, the entry section is stopped but reserve strip is fed from the entry accumulator to the furnace. After welding, the entry section drives are accelerated to a speed above the line speed until the entry accumulator is filled, and then decelerates to run at line speed.

It will be noted that the line includes five bridles which are used at points where it is necessary to alter the strip tension. Each bridle has its own drive and since the purpose of the rolls is to provide a constant differential tension between one point and another, the control is similar to that for a reel but is simplified by the roll radius and inertia remaining constant.

Each bridle has two driven rolls which must share the tension loads. To ensure correct sharing each roll is powered by a d.c. compound motor with cumulative and differential series fields (see page 52).

The bridle rolls move the strip through the process sections in conjunction with three powered furnace rolls (helpers) driven by d.c. motors supplied in parallel with No. 3 bridle motor. The speeds of the bridle drives must be set differentially to provide the tension required in the process section between bridles, and the relative speeds must be maintained with any normal alterations in the line operating conditions. The speeds are held to a regulation better than $\pm 0.1\%$.

The basic speed-setting control system is shown in *Figure* 8.7. For control of the line there is a master regulator with which the operator sets the drive speed level to obtain the strip speed required through the furnace and the pickling process. The speed level is set as a voltage, selected by adjustment of a potentiometer, which provides a speed reference signal to the speed regulators of Nos. 2, 3 and 4 bridles. No. 3 bridle acts as the line master running at the preset line speed. The speeds of Nos. 2 and 4 bridles are adjusted relative to that of No. 3 bridle to allow for tension control of the strip in the annealing furnace and pickle sections respectively. In the pickle section the relative speeds of the Nos. 3 and 4 bridles are such that the strip runs as a loop. Nos. 2 and 4 bridle speed regulators receive a tension-control signal in addition to the line speed reference signal.

The entry and exit sections of the line receive a speed reference signal from a tachogenerator (TG) driven by the master No. 3 bridle. These two sections also receive an overspeed signal to speed up Nos. 1 and 5 bridles when a reserve of strip has to be built up in the accumulators after a coil change. Derived from the position of the strip in the accumulators, the overspeed signal calls for an increase of 25% above maximum full speed. To facilitate starting after coil changes, acceleration of the bridle drives is controlled by a signal from a static ramp generator (see page 70) into the speed regulator of each bridle.

No. 1 bridle receives its speed reference signal from the No. 3 bridle

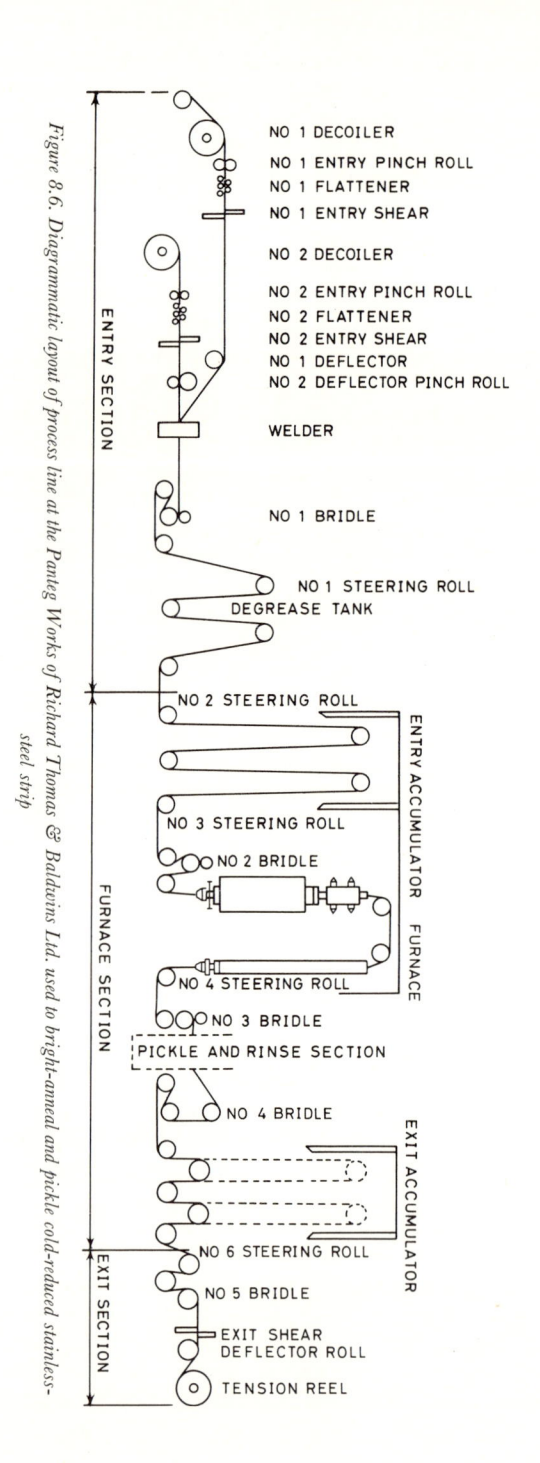

Figure 8.6. Diagrammatic layout of process line at the Panteg Works of Richard Thomas & Baldwins Ltd, used to bright-anneal and pickle cold-reduced stainless-steel strip

tachogenerator and in turn drives a tachogenerator providing a strip-speed signal to the c.e.m.f. regulator of the decoiler which controls the decoiler rotational speed in conjunction with the current regulator, in order to maintain the required tension throughout the run-down of the coil diameter and during changes in line speed. For tension control, the decoiler works against the No. 1 bridle in accordance with the tension reference signal, the speed-rate signal and the inertia compensation signal which is adjusted as the coil, or effective reel diameter, diminishes. When the No. 1 bridle receives its overspeed signal after being stopped for welding the strip ends together, the

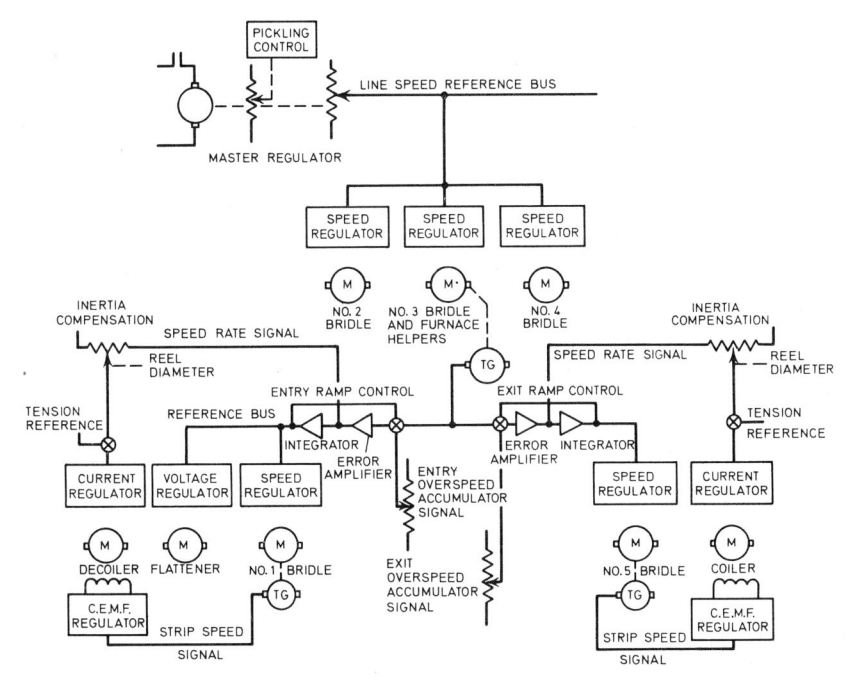

Figure 8.7. Basic speed-control system

control signals ensure that the decoiler accelerates at the rate necessary to maintain the set tension.

The flattener unit associated with each decoiler section (*Figure* 8.6) is used only up to a limited speed range to remove the coil set from the heavier gauge material and is controlled by a voltage regulator.

Running at overspeed No. 1 bridle refills the accumulator with enough strip to keep the furnace section operating at full speed for three minutes when the entry section is stopped again.

The tension of the strip through the furnace is controlled by the signal to the speed regulator of No. 2 bridle, this signal resulting from the difference between a voltage proportional to the tension required and a voltage proportional to the actual tension in the strip as measured by a load cell. The error signal offsets the bridle speed reference to increase or reduce the speed as required to adjust the strip tension.

The No. 4 bridle is controlled in a similar way to maintain the required strip catenary in the pickle section. The speed offset signal is derived from a load cell responding to alterations in the catenary.

The exit section storage accumulator is empty during normal running of the line but fills when No. 5 bridle and the coiler are stopped for removing a full mandrel and placing an empty one in position.

No. 5 bridle serves to anchor the strip while it is being sheared to enable the full mandrel to be removed. When the empty mandrel is in position, No. 5 bridle is started and accelerated to the overspeed value to empty the exit accumulator.

The No. 5 bridle and coiler are controlled as a unit in the same way as the No. 1 bridle and the decoiler but with the coiler functioning as a tension reeler with the coil diameter building up.

Each of the two driven bridle rolls is powered by a 385 V d.c. motor with a speed range of 0–1000 rev/min. The bridle roll motor ratings are: No. 1, 5·6 kW (7·5 hp) and 7·46 kW (10 hp); No. 2, 5·6 kW (7·5 hp) and 3·7 kW (5 hp); No. 3, 3·7 kW (5 hp) and 5·6 kW (7·5 hp); No. 4, 5·6 kW (7·5 hp) and 5·6 kW (7·5 hp); No. 5, 7·46 kW (10 hp) and 14·92 kW (20 hp). The furnace top roll and graphite roll motors are similar types with a maximum rating of 0·746 kW (1 hp), and they are connected in parallel with the No. 3 bridle roll motors.

Overspeed running of the entry and exit section drives is obtained by increasing the armature voltages to 490 V maximum. The basic bridle speed regulator shown in *Figure* 8.8 regulates the armature voltage, with fixed excitation of the shunt field. The armature supply is provided by a figure-8 thyristor-converter system (see page 62). The speed regulator contains an inner current control loop, which is identical to that in the current regulator used for the decoiler and coiler drives.

The speed signal inputs to amplifier A_1 produce an output signal to the current-control amplifier A_2, which produces a current-error signal to amplifiers A_3 and A_4 controlling the firing angle of the associated thyristor converter, which is held out of conduction by a biasing voltage. A speed feedback signal is obtained from the motor-driven tachogenerator.

The speed signal error determines the value of armature current required to provide the drive torque but with the inner current-control loop the maximum excursions of the signal are limited so that the maximum current is restricted to a safe value. The rate of change of current is also limited to the value necessary to avoid motor commutation problems, and the improved response to load changes obtained with the current-control loop assists speed recovery with step-load changes.

The current control loop holds the armature current to an accuracy of 1·5% over the full speed range and during acceleration and deceleration by regenerative braking. There is a current feedback to amplifier A_1, the maximum output of which is limited by adjustable amplifier-output limiters used to set the current-limit levels.

The decoilers are powered by 385 V drag-generators rated at 0/15/15 kW at 0/500/1500 rev/min, and the coiler drive motor is rated at 0/33·5/33·5 kW (0/45/45 hp) over the same speed range. Basically the coiler and decoiler have similar control functions, i.e. to keep the coil under a constant tension during winding or unwinding. Winding tension control is in accordance with the principle that, if the motor back e.m.f. is kept proportional to strip

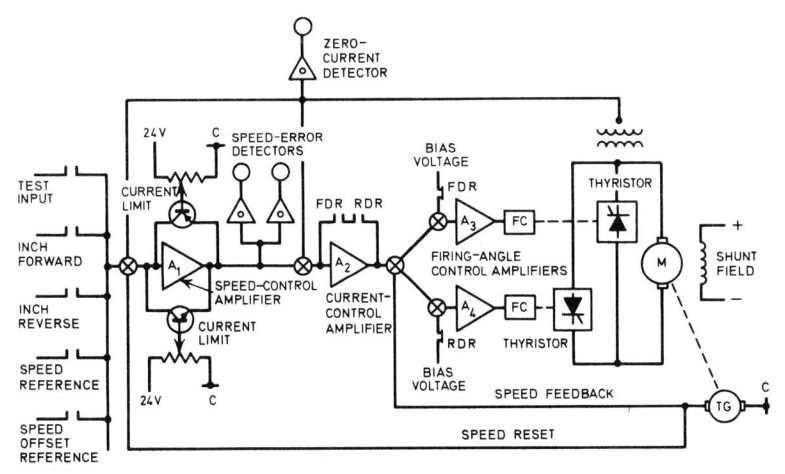

Figure 8.8. Bridle speed regulator

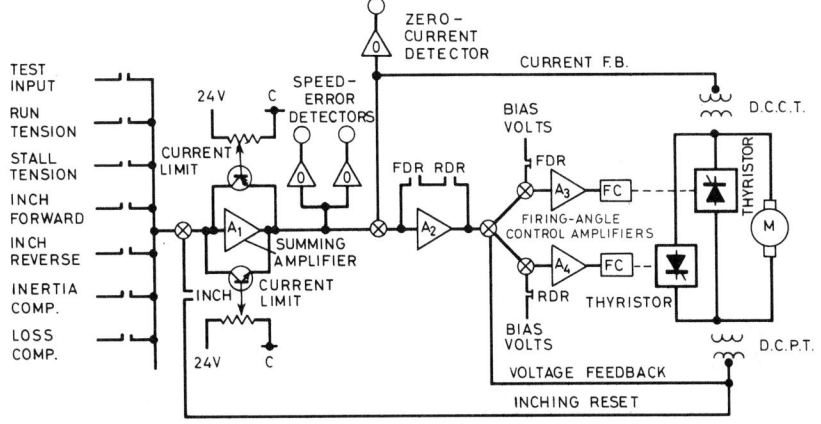

Figure 8.9. Current regulator

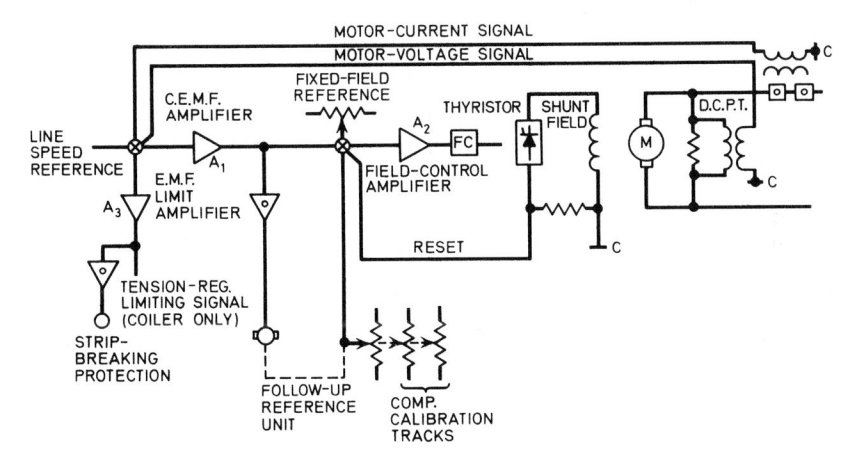

Figure 8.10. Counter e.m.f. regulator

speed, the motor armature current will be proportional to strip tension. The same principle applies to the decoiler drag-generators but in this case it is the generated e.m.f. that is kept proportional to strip speed as the coil radius decreases. Each decoiler and the coiler are controlled by a current regulator (*Figure* 8.9) and a so-called counter e.m.f. regulator (*Figure* 8.10), which controls the thyristor converter supplying the shunt field winding of the drive machine.

The current regulator (*Figure* 8.9) maintains the motor armature current in accordance with the required tension set by the operator. Inertia and drive loss compensation signals modify the current reference to keep the strip tension constant and independent of strip speed.

The tension, inertia and loss compensation signal inputs to amplifier A_1 produce an output to amplifier A_2, which sums the current reference and the current feedback signal from the d.c. current transformer and amplifies the current-error signal.

The current-regulator characteristics will vary over the speed range as a function of the change in the motor back e.m.f. Since the current regulator is used as a tension controller, the current-reponse characteristic is kept sensibly constant by compensation from motor voltage, thus keeping the loop gain constant over the speed range. A voltage feedback signal is obtained from a d.c. potential transformer.

If the strip being coiled breaks so that tension is lost, the current regulator functions to raise the speed of the coiler. To limit the rise in speed the c.e.m.f. regulator functions to maintain the motor speed slightly above the value proportional to normal strip speed. Motor speed is monitored continuously by comparing the line speed reference with the motor e.m.f. via amplifier A_3 (*Figure* 8.10) which gives an output only if the motor e.m.f. exceeds that appropriate to the immediate coiling conditions. An output is first fed to the current regulator to cancel the tension control and hold the motor speed, and secondly it stops the motor by a signal to the strip-breakage protection circuit.

The c.e.m.f. regulator has two control loops based on the amplifiers A_1 and A_2 respectively. The latter controls the field supply thyristor converter by an output determined by three inputs. In the case of the coiler one input is the weak-field reference while for the decoiler the input is a full-field reference. The second input is the c.e.m.f. error signal from amplifier A_1 which boosts the coiler weak-field reference and bucks the decoiler full-field reference as required for field-strength variation for speed regulation. The third input is the signal from the follow-up reference unit.

This unit is a servo-driven rheostat which follows-up the change in the c.e.m.f. error signal as the coil radius alters and the motor speed varies to keep the required strip linear speed constant. The rheostat also has compensation calibration tracks providing the current regulator with the inertia and loss compensation reference signals which change as the coil radius and motor speed change.

The c.e.m.f. amplifier A_1 gives only a buck or boost signal output which is limited to provide the appropriate field excitation limitation required. The input signal is obtained by comparing the line speed reference with a motor-voltage signal from a d.c. potential transformer.

Bibliography

GLEDHILL, J. T. 'Blast Furnace Charging Equipment for the Workington Iron and Steel Company', *AEI Engineering*, September/October (1962).
READ, J. C. 'Rectifiers and Rectifier Applications', *Proceedings IEE*, **110,** No. 4, April (1963).
BAINES, A. P. 'The Control of Mercury-arc Convertors for Industrial Drives', *English Electric Journal*, July/August (1963).
PETTIT, R. 'Flux-reset Magnastats', *AEI Engineering*, January/February (1963).
EALES, L. W. 'Standardised Variable-speed Drives Using Semiconductor Devices', *GEC Journal of Science and Technology*, **31,** No. 1, (1964).
STRETCH, M. B. K. 'A Statically Controlled Finishing Mill for Aluminium Strip', *AEI Engineering*, September/October (1964).
VARLEY, K. 'An Improved Contactless Controller for Planer Drives', *AEI Engineering*, November/December (1964).
TOFT, J. G. 'The Ward-Leonard Friction Drive at Manton Colliery', *AEI Engineering*, January/February (1965).
SCHWARZ, K. K. 'Automatic Control of N-S Variable-speed A.C. Motors', Publication 129, Laurence, Scott & Electromotors Ltd.
BONE, J. C. H. and SCHWARZ, K. K. 'Automatic Control of Variable-speed Pumps', Publication 165, Laurence, Scott & Electromotors Ltd.
SIMONS, D. S. 'Reversing Hot Mills and Their Auxiliaries', *AEI Engineering,* January (1967) (Electrical Engineering in the Metal Industries supplement).
PEPWORTH, F. J. and SKENFIELD, B. 'D.C. Motors for the Metal Industries, *AEI Engineering*', January (1967) (Electrical Engineering in the Metal Industries supplement).
YOUNG, W. J. 'Continuous Hot Mills', *AEI Engineering*, January (1967) (Electrical Engineering in the Metal Industries supplement).
PECK, P. E. and RAYDEN, R. G. 'Special Mills' *AEI Engineering,* January (1967) (Electrical Engineering in the Metal Industries supplement).
ROBERTS, L. J. 'Process Lines—Special Problems', *AEI Engineering,* January (1967) (Electrical Engineering in the Metal Industries supplement).
JONES, R. R. and DUGARD, N. A. 'Bi-directional Thyristor Drives for a Steelworks Process Line', *GEC Journal*, **34,** No. 3 (1967).
WILSON, R. R. 'Electric-motor-operated Actuators', *LSE Engineering Bulletin*, **9,** No. 3, February (1967).
MUSHAM, R. 'Reversing Static Contactor', *Electrical Times*, May 25, June 8, (1967).
RODWELL, R. G. 'Thyristor Convertors for Large Drives', *Electrical Times*, August 3, (1967)
GREENWOOD, P. B. 'Synchronous Reluctance Motors', *Electrical Review*, March 22, (1967).
DRAPER, D. W. and GOODRIDGE, R. T. 'Thyristor-supplied Tandem Cold Mill', *Proceedings IEE*, **115,** No. 10, October (1968).
BONE, J. C. H. 'Variable-frequency Drives', *Electrical Review*, October 18, (1968)
BAILEY, B. G. 'Roller Table Drives in Heavy Plate Mills', *LSE Engineering Bulletin*, **10,** No. 3, March (1969).
FLACK, R. F., GRANT, W. T. and PARKER, C. 'Applications for Eddy Current Couplings', *Electrical Times*, April 17 (1969).
SMITH, J. W. 'Closed-loop Control of High-performance Variable-speed Drives', *LSE Engineering Bulletin*, **10,** No. 4, October (1969).

Index

A.C. thyristor contactors, 21, 23, 25
Acceleration and deceleration control, 3, 7, 70
Air-operated reversing switch, 120
Aluminium foil mill drives, 108
Aluminium strip finishing mill drive, 121
Aluminium tension levelling line bridle drive, 74
Amplidyne exciters, 54, 107
Amplifiers, control:
  amplidyne rotating, 54, 107
  function of, 2, 4
  Kom-Pac set-point, 86
  magnastat, 82, 121
  magnetic, 82, 97
  transistor, 84
Analogue speed reference voltages, 70
Anti-parallel connected thyristor bridges, 63
Ascron thyristor switching system, 25, 87, 97
Asmag control system, 97
Asrec control system, 93
Automatic gauge control, 76, 112
Automatic on/off control circuits, 66
Automatic on/off switching devices, 66
Auxiliary compole windings, 51

Back e.m.f. speed regulators, 75, 131
Back e.m.f. speed signals, 72
Basic d.c. motor speed control system, 2
Billet mill stand d.c. motor drive, 82
Booster generators in Ward-Leonard control schemes, 55, 74, 102
Bridle drives by d.c. compound motors, 52, 129
Bridle speed regulators, 74, 129, 132

Closed-loop systems, 4
Coil-radius servo system, 116
Coilers (see Reeler drive control, rewind)

Commutating poles, auxiliary, 51
Commutation conditions, 7, 49
Compound wound d.c. motors (see D.C. compound wound motors)
  bridle drives, 52, 129
  cumulative and differential fields, 52, 129
Constant cutting speed scheme, 95
Constant speed control systems, 95
Contactless switches, 68
Continuous hot rolling mill control requirements, 17
Continuous process line control requirements, 9
Continuous speed control schemes, 69
Control amplifier signal circuits, 85
Control amplifiers, 2, 4, 54, 82, 84, 86, 97, 107, 121
Control signals:
  sources of, 5
  utilisation of, 4
Controlled quantity measuring devices, 73
Converter-fed reversing drives:
  armature-reversal, 59
  control requirements, 57
  double-converter, 59
  field-reversal, 59
Conveyer control scheme, 88
Counter e.m.f. d.c. motor regulator, 131, 134
Crab-mounted bucket-hoist drive, 107
Current-limit control, 3, 6
Current regulator for tension control, 75, 132, 134
Cyclo-converter, 31

D.C. compound wound motors:
  bridle drives, 52, 129
  cumulative and differential fields, 52, 129
D.C. injection braking of induction motors, 27
D.C. link-type static converter, 31

138 Index

D.C. series wound motor, 51
D.C. shunt wound motors:
  auxiliary compole windings for, 51
  commutation requirements, 49
  converter-fed reversing drive, 58
  exciters, 49
  for reversing duties, 49
  for variable-speed drives, 48
  power regulators for, 53
  torque/speed characteristics, 47
Dancer-roll loop control, 77
Dancer-roll operated potentiometers, 78, 126
Dancing-roll operated variable reactor, 77
Decoilers (see Reeler drive control, unwind)
Diode rectifier-fed d.c. motor drives, 63
Diode rectifier power regulators, 63
Double-converter reversing drives, 128
Drag (brake) generator, 13, 119, 132
Drive acceleration and deceleration control, 3, 7, 70
Duty cycle control requirements, 6
Dynamic braking d.c. motors, 83, 123

Eddy-current brake, 37
Eddy-current coupling, 34
Electronic ramp-function generator, 70
Exciters, amplidyne, 54, 107

Float-operated potentiometer, 95
Flow sensors, 73
Flux-reset magnastat amplifier, 82, 121
Four-roll bridle drive, 52
Frequency converters, 30
Fully controlled thyristor bridge, 62

Group motor starting, 139

IR armature voltage drop compensation, 72
Ilgner drive induction motor slip control, 34
Impact speed drop, 17
Induction motor drives, 27, 34, 86
Induction regulators for N–S type motors, 39
Inertia compensation, reeler, 13, 76, 118
Inter-bridle speed ratio control, 74
Interlocking control schemes, 66

Level control schemes, 86
Level sensing electrodes, 86
Limit switches, 66, 69
Load cells, 78
Logic functions, 68
Logicon static switching system, 121
Loop position control, 14, 77, 79, 126

Magnastats in control schemes, 82, 121
Magnetic amplifier control of shunt motor field, 82

Manual control circuits, 66
Mercury-arc converter:
  operation as power regulator, 55
  operation for inversion, 58
  reversing drives, 120
Mine winder:
  control characteristics, 70
  duty cycle, 7
Motor loss compensation, reeler, 76, 114
Motor-operated preset speed controllers, 99
Motorised potentiometers, 111, 127
Multi-motor variable-speed pump drives, 96
Multi-speed squirrel-cage motors, 27

N–S type 3-phase commutator motors, 38, 95
Noflote level control scheme, 86

On/off control circuits, 66
Open-loop control, 4

Pay-off reelers (see Reeler drive control, unwind)
Photoelectric loop position control schemes:
  for free-hanging loop, 79
  for upward loop, 81
Planer table drive, 106
Plug braking (plugging), 27
Pneumatic coil-radius measuring device, 117
Pole-amplitude modulated (PAM) motors, 28
Preset speed controllers for Schrage type motors, 98
Process line strip accumulators, 129, 131
Pulse generators for speed signals, 71, 73
Pump drives:
  level control of, 86, 95
  multi-motor, 96
  single-motor, 86
  variable-speed, 95

Ramp-function generators, 70, 121, 129
Reeler drive control:
  centre-spindle type, 12, 13, 75
  constant-power, 105
  drag (brake) generators for, 13, 119, 132
  inertia compensation, 13, 76, 118
  loss compensation, 13, 76, 114
  mechanical brakes for, 13, 15
  rewind, 9, 11, 12, 75, 108, 113, 132
  surface-type, 11
  tension tapering, 12, 78
  typical schemes, 32, 75, 102, 104, 113, 129
  unwind, 9, 10, 13, 108, 113, 129, 131
Reeler types:
  centre-spindle, 10
  surface, 9
Reeling systems:
  free, 11

Reeling systems *continued*:
  tension, 11
Refrigeration plant control scheme, 90
Regenerative braking, 54, 58, 119
Relsyn synchronous reluctance motors, 32
Resistance-capacitance stability circuits, 107
Reversing hot-rolling mill, 8
Roller table drives, 31
Rotary frequency converters, 30
Rotary multiplier, 105

Saturable reactor (transducter) power regulator, 65
Schrage-type 3-phase commutator motors, 43, 98
Semiconductor logic gates, 68
Sequential motor-starting schemes, 87
Shunt field bias windings, 54, 84
Shunt wound d.c. motors (see D.C. shunt wound motors)
Signal mixing magnastat, 82, 121
Slipring induction motor drives, 34, 87
Speed control signal devices:
  back e.m.f. bridge, 72
  controlled parameter sensors, 73
  float-operated potentiometer, 95
  pulse generators, 71
  tachogenerators, 71
  rotary multiplier, 105
Speed reference voltage setting devices:
  decade switches, 70
  motor-driven potentiometers, 70
  potentiometers, 69
Speed regulators:
  d.c. shunt field, 53
  frequency converter, 30
  induction, 39
  mercury-arc converter, 55
  preset controller, 99
  rectifier-fed d.c. field, 63
  rectifier-transformer tap-changer, 63
  thyristor converter, 58
  Ward-Leonard, 53
Spillover speed control technique, 17, 77, 111, 123
Squirrel-cage motor drives, 27, 28, 86
Stabilised reference voltage units, 69
Stardrive thyristor converter units, 123
Start/stop duty control requirements, 6
Static frequency converters, 31
Static switching systems, 66
Stator-fed 3-phase commutator motors:
  control systems for, 93
  speed regulation of, 37
Steel strip-processing line drives, 128
Strip coiler control scheme, 75
Strip mill loop control, 14, 16, 77, 79, 126
Strip tension control:
  by armature current regulation, 75
  by load cell signal, 77

Strip tension control *continued*:
  by speed signals, 74
  by tensiometer signal, 74
Suicide fields, 54
Synchronous-reluctance motor, 32

Taper tension control, 78
Tachogenerators, 71
Tektor level controller, 86
Temperature sensors, 73
Textile machine drives, 96
Thyristor,
  Ascron switching system, 25, 87, 97
  burst-firing (integral action) control, 25
  phase-shift (phase-angle) firing control, 24
  power regulators, 24
  switching action on a.c., 22
  switching action on d.c., 22
Thyristor contactors, 22, 23
Thyristor converter-fed reversing drives, 58, 124
Thyristor/diode mixed bridge, 62
Toothed-wheel pulse generator, 71
Torque motor drives, 32
Transducers, 73
Transductor power regulator, 65
Transistor amplifier, 84
Two-stand cold rolling mill, 14
Type CH 3-phase commutator motors, 44, 98
Tyre-bead production plant drives, 127
Tyre tread-section production line drives, 124

Unwind reelers (see Reeler drive control, unwind)

Valve actuator drives, 86
Variable-frequency induction motor drives, 28
Variable-frequency roller-table motors, 31
Variable-frequency supply equipment, 28
Variable-frequency synchronous motor drives, 32
Variable-speed coupling, 34
Variable-speed pump drives, 95
Variable-speed 3-phase commutator motors:
  Schrage-type, 44, 98
  stator-fed type, 37, 93
Velonic Ward-Leonard scheme, 102

Ward-Leonard drives, 7, 34, 49, 53, 70, 74, 102
Ward-Leonard set equipment:
  booster-generator voltage regulator, 55
  excitation system, 49, 54
  generator-field thyristor regulators, 55
  Ilgner induction motor-generator, 34